WORKING THROUGH SURVEILLANCE
AND TECHNICAL COMMUNICATION

SUNY series, Studies in Technical Communication
———————
Miles A. Kimball, Charles H. Sides, Derek G. Ross,
and Hilary A. Sarat-St. Peter, editors

WORKING THROUGH SURVEILLANCE AND TECHNICAL COMMUNICATION
CONCEPTS AND CONNECTIONS

SARAH YOUNG

Cover Credit: *Fearless Girl* by Kristen Visbal.

Published by State University of New York Press, Albany

Open Access funded by Erasmus University Rotterdam Library in support of open science initiatives.

The text of this book is licensed under a Creative Commons Attribution-NonCommercial-NoDerivatives 4.0 International License (CC BY-NC-ND 4.0).

© 2023 State University of New York

All rights reserved

Printed in the United States of America

For information, contact State University of New York Press, Albany, NY
www.sunypress.edu

Library of Congress Cataloging-in-Publication Data

Name: Young, Sarah, 1980– author.
Title: Working through surveillance and technical communication : concepts and connections / Sarah Young.
Description: Albany, NY : State University of New York Press, [2023] | Series: SUNY series, studies in technical communication | Includes bibliographical references and index.
Identifiers: LCCN 2022035145 | ISBN 9781438492766 (hardcover : alk. paper) | ISBN 9781438492773 (ebook) | ISBN 9781438492759 (pbk. : alk. paper)
Subjects: LCSH: Electronic surveillance—Moral and ethical aspects. | Data privacy. | Technical communication.
Classification: LCC TK7882.E2 Y68 2023 | DDC 621.389/28—dc23/eng/20221018
LC record available at https://lccn.loc.gov/2022035145

10 9 8 7 6 5 4 3 2 1

*This is for my mother who passed away
while we were separated by the
Atlantic and the pandemic.
From the moment I left your driveway,
I didn't mean to be gone for so long.*

Contents

List of Illustrations — ix

Acknowledgments — xi

Introduction — 1

1 Introduction to Surveillance and Technical Communication — 5

2 Surveillance Workers and Technical Communicators — 41

3 Information, Technical Communication, and Surveillance — 61

4 Evaluations and Responses: Social Justice, Ethics, and Surveillance — 83

5 Resisting Surveillance through Tactical Communication and Social Justice — 117

6 Surveillance Writing: A Pedagogy — 133

Conclusion — 159

Notes — 165

References — 171

Index — 201

Illustrations

Figures

Figure 2.1	A Visual of the Range of the Surveillance Worker	46
Figure 2.2	A Visual of the Range of the Technical Communicator	50
Figure 2.3	A Matrix of Technical Communicators and Surveillance Workers	51

Tables

Table 3.1	Sample of NSA-Associated Surveillance Disclosed by Snowden	65
Table 3.2	Sample Surveillance Paradigms	70
Table 3.3	Elements in a Surveillance Scenario	73
Table 3.4	Analysis of Performance Review Report	76
Table 3.5	Analysis of a Personal, Professional Web Page	77
Table 3.6	Analysis of Institutional Website	78
Table 3.7	Social Media User Experience Analysis	80
Table 4.1	Ethics, Snowden, and the NSA	103
Table 6.1	Genres of Surveillance Writing	139
Table 6.2	Sample Course Schedule for Surveillance Writing	147

Acknowledgments

Thank you to my closest supporters, Chad, Baxter, and Wilder, and my family, as well as my four-legged friends who have come and gone during the writing of this book.

Professional thanks to mentors Jason Pridmore, Daniel Trottier, Peter Goggin, Greg Wise, Alice Daer, Catherine Brooks, and Maryfrances Wagner for their encouragement over the years. Also, thanks to the anonymous reviewers who stuck with the various iterations of the draft; the editors of this series, Miles A. Kimball, Charles H. Sides, Derek G. Ross, and Hilary A. Sarat-St. Peter for their support in this project; the MAPS group at Erasmus University Rotterdam; and thanks to Sam Dragga for bringing issues like the Immigration and Customs Enforcement memo discussed in chapter 6 to listserv attention.

Early drafted portions of this book were completed during a LEaDing Fellows fellowship at Erasmus University Rotterdam, which received funding from the European Union's Horizon 2020 research and innovation program under Marie Skłodowska-Curie grant agreement no. 707404.

Introduction

Surveillance matters. But it is complicated. It is both ever-present and ubiquitous, often provoking strong responses, but also fuzzy and gray. Surveillance is best viewed at localized, empirical levels to define, measure, evaluate, and respond to it. But one cannot do those things unless one knows what to look for. This book aims to reach those in the disciplines of technical communication (TC) and surveillance, or anyone who might find themselves using technology or technical information to communicate information to help recognize, assess, and find agency to resist spaces of surveillance if necessary.

In the past ten years, nothing illustrates the complexities of surveillance better than the case of Edward Snowden. A contractor for the United States' National Security Agency (NSA), in June 2013, Snowden gave journalists classified information detailing the United States' mass surveillance systems throughout the world. At the time of the release of information, I was an investigator in the background investigation industry, an industry that vetted Snowden and gave him access to this information. While I was not involved with Snowden's investigation, his actions had repercussions for me. The background investigation industry went into a tailspin trying to figure out how Snowden could have "slipped through the cracks." Now an investigator-turned scholar, I am trying to make sense of the situation, assessing the dimensions of the Snowden case, and finding overlaps between both surveillance studies and TC.

But this book is both all about and nothing about Snowden. Snowden represents someone who was known to work in surveillance and the intelligence community, but he also represents a technical communicator who carried out duties of surveillance. Whether one agrees with his actions or not, in the abstract, he represents someone who had supposedly evaluated his role and the ethics of a scenario of surveillance and decided not only to end his involvement but also to broadcast what he knew about surveillance to the public, arguing for more transparent systems. It is through his experience that we see a range of what it means to be a technical communicator and a surveillance worker, how definitions of both are simultaneously evolving, and how ethical and socially just decision-making requires being sensitized to the idea of surveillance in the first place. Further, being sensitized to the dimensions of surveillance are crucial for reacting to it, teaching about it, and for understanding surveillance futures. Surveillance can be a heuristic for those working for social justice.

At this book's publication, it has been almost ten years since Snowden helped disclose classified information that brought visibility to invisible systems. But instead of being more empowered, as time has passed, there is evidence that some feel more powerless, especially in regard to online surveillance (Boerman et al., 2021). These feelings are particularly important, because when individuals don't feel like they have efficacy, "people will usually not establish protecting behavior if they do not believe that this behavior is effective" (p. 968).

Protection of one's information isn't the only way to address or feel empowered against surveillance, however, and the goal of this book is to address other ways to think about consequences of surveillance beyond "privacy harms." Privacy harms put the onus of resistance on the individual to stay private and emphasize the idea that surveillance is consistently bad. But the consequences of surveillance tend to be not only a moving target; they can also be simultaneously beneficial. For instance, are TikTok content producers "harmed" by algorithms? There could be direct threats from stalkers, threats of violence, or chilling effects where a producer decides against posting something for fear of some other social consequence. But there can also be enjoyment, wanted attention, and needed socializing and fun. So, perhaps yes, perhaps no.

In this complex space, *Working Through Surveillance and Technical Communication: Concepts and Connections* explores the ongoing conversations between TC and surveillance using Snowden's high-profile example. It enters in at conversations of labor, employment, watching, visibility, ethics, social justice, media, order, and resistance, all through the lens of surveillance. Often, privacy is framed as the way to reduce the impact of surveillance, but instead, this book places more focus on how evaluating and reducing surveillance can also reduce the impact of surveillance, particularly when practices are oppressive. Focusing on reducing surveillance thus puts the agency and potential for efficacy back on the technical communicator who might be working with (surveillance-enhancing) technologies or otherwise watching someone else or their data, anonymously or not.

Taking a more inward look at this book, its overall purpose then is to keep this backstory in mind and explore connections between TC and surveillance studies a little more, in order to offer disciplinary entry points to both surveillance and TC. The book is led by the central question, how are technical communicators also surveillance workers, and why does this matter for TC and surveillance scholarship? To do this, each chapter will mix disciplinary conversations about TC and surveillance studies, based on the theme of each chapter, for an interdisciplinary conversation. That said, this book will not attempt to exhaustively explicate either fields of TC or surveillance studies and does not attempt to fully explore all dimensions of the illustrative example of Edward Snowden. It will, however, be a survey and offer entry points into a larger surveillance conversation where TC and other scholars have already been exploring but still have room to expand.

Chapter One

Introduction to Surveillance and Technical Communication

In June 2013, media installations such as *The Guardian* and *The Washington Post* published classified United States government information provided to them by NSA contractor Edward Snowden. This information detailed surveillance programs, collaborations, and databases that the US was involved with, both nationally and globally, like the now well-criticized programs of PRISM, Five Eyes, or XKeyscore, and the Enterprise Knowledge System (as will be discussed more in chapter 4).

After Snowden's revelations, there were debates as to whether he was a traitor or a whistleblower (Boehme, 2018) and debates over the limits of surveillance capabilities (e.g., Cassidy, 2013; Page, 2015). While the US justified surveillance programs by saying surveillance was necessary for homeland and national security purposes (Campbell, 2013), the revelations led others to question the government's abilities to legitimately collect information en masse such as telephone records, metadata, online data, and information about world leaders and Americans, especially those not currently under suspicion of crimes. That mass surveillance was being conducted by the US government on all sorts of people, beyond terrorists or enemies of the state, was particularly concerning (Lyon, 2015) and incited global protests (Gabbatt, 2014; Reuters, 2013; *South China Morning Post*, 2013).

While no consensus among governments, corporations, and publics is likely to be achieved any time soon as to the limits of surveillance practices, expectations of privacy, or the classification of Snowden's actions, Snowden's actions made the public more keyed into and resistant to dominant rhetorics of surveillance. Snowden commented confidently on the growing awareness, in retrospect, six years later when he noted with hope, "The government and corporate sector preyed on our ignorance. But now we know. People are aware now. People are still powerless to stop it but we are trying. The revelations made the fight more even" (MacAskill & Hern, 2018, para. 5).

Despite the increasing awareness, though, as noted, it has been nearly ten years since Snowden's disclosures, and it becomes useful to evaluate and reflect on how a disciplines and discourses like TC have used surveillance in the past as well as reflect on where conversations can go in the future. As will be discussed later in this chapter, TC scholars have long been looking at surveillance themes, but with a proclivity toward ethics and a growing focus on social impacts, surveillance also offers a fruitful lens to use to pull out even more critiques of, say, technological access, visibility, or the right to not be watched or classified.

One way to do this is by pulling on research from surveillance studies. It is particularly useful to do so, because, as Bowker and Star (1999) bring out, "Good, usable systems disappear almost by definition. The easier they are to use, the harder they are to see" (p. 33). Fields of study become systems of information, organizing, limiting, and defining how questions should be asked and interpretations that can be made. Further, the languages in the fields also become terministic screens, or lenses that we use to make sense of the world. Burke's (1990) popular quote speaks to similar concepts when he says, "[M]uch that we take as observations about 'reality' may be but the spinning out of possibilities implicit in our particular choice of terms" (p. 1035). For Burke, the terministic screen is the way that terms both define what is meant and what is not meant at the same time. To say something *is* something means that it is not the other. So, inquiries comparing two fields about the same topic are useful in bringing together different discourses that often speak about similar things.

With this preface as a backdrop, as a brief sketch of what is to come, this book will evaluate the technical communicator and surveillance worker in both a formal and informal sense to explain the varying degrees to which one could be engaged in both practices of TC and/or surveillance (and both at the same time). An evaluation of roles will make it clearer how the concept of "information" connects the surveillance worker and communicator and shows how we must rethink traditional definitions of *surveillance* to see how a technical worker can carry out "everyday" practices of surveillance. The book also describes why surveillance is an important concept by looking at issues of ethics and social justice; then it provides examples of how and why to resist problematic surveillance. It closes by offering ways to use the Snowden example to teach and explore rhetorics of surveillance. Teaching about surveillance is particularly important due to the complexity of surveillance itself. As repeated throughout the book, surveillance is not only definitionally a "contested concept" that Gallie (1956) and Mulligan (2016) describe (which is a term that creates an endless dispute about meaning), but it is also very contextual. A watchful eye over some may be acceptable to one group at any given time in a certain place, but the same surveillance could be problematic to others under different (or even the same) conditions. Students who don't see surveillance as a "sensitizing concept" (Blumer, 1954), which they can use for a "general sense of reference and guidance" of the world (p. 7) may overlook spaces to critique surveillance and its rhetoric. If one isn't sensitized to what these terms might be in the first place, one may not "see surveillantly" (Finn, 2012, p. 67) which could mean they don't recognize that they themselves are participants in practices of surveillance.

Further, this book can serve as constitutive rhetoric in so far that those not typically considering themselves in either TC or surveillance communities, especially at the same time, can find themselves in the discourse of either or both roles. To do this, each chapter will introduce concepts from both surveillance studies and TC and then use the illustrative example of Edward Snowden to make the points clearer and more relevant to both contemporary surveillance and TC practices. Finding oneself in the surveillance discourses can lead to more responsible, ethical, and empowering uses of technology and technical information at the juncture of surveillance practices.

An Overview of Studies of Surveillance

In that context, it is useful to start this conversation with a discussion of surveillance. Marx (2005) comments that "traditional" definitions of surveillance found in the dictionary talk about criminals, and these definitions often sound similar to the "close observation, especially of a suspected person" (p. 817). McQuade and Danielson's (2005) conventional definition of surveillance as the "targeted monitoring of activities by police or security officials for specific evidence of crimes or other wrongdoing" (p. 1,228) reflects this connection. These definitions correspond to Hollywood depictions of a shadowy FBI agent in a van watching a mark, or we can imagine a documentary reenacting the moments that a team of analysts stopped a cohort of terrorists plotting to carry out a nefarious scheme.

Dictionary definitions are not what this section is about, however, and dramatic scenes from Hollywood's surveillant imagination[1] fail to capture the breadth of what surveillance can entail. As already noted, surveillance is an intricate term that can't easily be boiled down to a pithy, all-encompassing statement. However, although surveillance is a contested concept, studies of surveillance provide a rich way to think about various possible entry points for exploring its dimensions.

While trying to classify and consider as much literature as possible is also a daunting task, there are at least two useful ways to go about presenting an overview of surveillance: using paradigms and summaries. Paradigms, particularly those from Foucault and Deleuze and the terms *surveillance capitalism* and *everyday surveillance* speak to larger patterns, and summaries provide insights at a more microlevel assessment of the themes frequently addressed by surveillance scholars.

Starting with paradigms makes the most sense, because one of the most dominating paradigms for surveillance research is Foucault's notions of panopticism and disciplinary power, and this paradigm also happens to have been one of the most visible ways of approaching surveillance for the past forty-five years (this will be evidenced later in the discussion of surveillance in technical communication research).

Michel Foucault (1977), as one of the foundational modern theorists of surveillance, used an eighteenth-century prison design of the panopticon by Jeremy Bentham to describe how disciplinary power is maintained by governmental and institutional monitoring, or what can be considered surveillance. As noted by Brunon-Ernst (2012), Bentham theorized several versions of the panopticon, all with the intention that any entity in a position of power could keep a central watch over everything else that surrounded the watcher, and that those subjected to the gaze of the powerful would conform to the expectations of those in power by succumbing to their lack of control over privacy. In a prison-type panopticon, a guard tower sits in the middle of a ring of jail cells. Prisoners in the cells can't see the guards inside the tower due to shaded/blinded windows, and supposedly, because the prisoners don't know when the guards are looking, they control and discipline themselves to conform to institutional expectations. When studying prisons, Foucault gravitated toward this prison panopticon design; he found it to be "the epitome of the transformations taking place at the time" (Brunon-Ernst, 2012, p. 17).

The interesting thing that Foucault does with the panopticon, though, is take the idea of watching away from exclusively state-sponsored justice to apply the panopticon to other situations with similar power structures, leading from the specific architectural Bentham "panopticon" to Foucault's idea of "panopticism" that could be used as a metaphor in different contexts. This repositioning became a dominant paradigm for Foucault to explain "surveillance as involving an all-seeing inspector" (Galič et al., 2017, p. 12). Foucault used the idea of panopticon to further develop his idea of the disciplinary society, which uses principles of panoptic power to explain that if one can "potentially be under surveillance, people will internalize control, morals and values" and thus this discipline becomes a tool of the more powerful (p. 16). Foucault postulated that this situation was like systems of visibility present in a variety of other institutions such as schools, hospitals, and workplaces. People within those organizations become docile bodies conforming to norms because someone might be watching.

Scholars across fields have come to use Foucault's metaphor and his idea of visibility for all sorts of explanations of watching and power, so much so that Kevin D. Haggerty (2006) notes, "For a quarter-century, the panopticon has been the exemplar for inquiries into surveillance," and panopticism, especially as applied to other institutions beyond a prison, has "emerged as one of the most popular concepts in contemporary social thought" (pp. 24–25).

However, Haggerty continues, because of this popularity, the panopticon has become an all-encompassing metaphor for surveillance, and this is not necessarily good for three reasons. First, as a metaphor, it ultimately can't account for or explain every instance of surveillance. Panopticism has been tried and forced to fit all sorts of situations, with the prefix altered so many times just to try to reflect more diverse meanings.[2] Ultimately, though, these alterations still go back to the basic metaphor, which has led to overlooking and undercriticizing more diverse spaces of surveillance that panoptic power doesn't smoothly explain. Second, when all the attention is turned toward panopticism, there is less attention toward other ways to explain surveillance, so when we stick to panopticism, we're also limiting our imaginations and how we can frame and explore surveillance in other ways. Finally, while Foucault is a foundational theorist in the way he describes surveillance practices and extends surveillance beyond the state, and while he made a foundational mark in studies of surveillance, more recently, scholars have pointed out that there are limitations to Foucault's framing, particularly due to technological advances, and that surveillance is no longer limited to brick-and-mortar locations with physical bodies like those so frequently equated with panopticism. Often contemporary surveillance has become ubiquitous (Allen, 2019; Andrejevic, 2012),[3] as well as a routine practice built into infrastructure of governments, technologies, companies, and the home and gathered by a variety of hosts in dispersed, rhizomatic ways that pop up in many spots (Haggerty & Ericson, 2000).

This shift has led scholars to new directions. While, on the one hand, Foucault and his work is still valuable and still circulates, on the other hand, another competing paradigm has emerged in popularity. This is Gilles Deleuze's (1992) theories of control. Although not divorced from Foucault's work, Deleuze postulates that in the control society, control is not coming from specific locations such as guard towers or institutions, but rather control is constant because of technological affordances that can track us everywhere. In a control society, one never leaves being monitored, and one is constantly being scrutinized. As opposed to Foucault and the disciplinary society's strong emphasis on institutions in brick-and-mortar locations with physical watching, the control society thesis is less about embracing ideas of central stakeholders, unmediated by technology in fixed places. The model is more about maneuvering through a culture of control where surveillance is embedded in our daily actions through technology. Corporations, financial systems, social media sites, and loyalty cards—these can all be sites of surveillance with surveillers who can keep track of individuals beyond their actual physical walls (if those even exist) through technological means.

Further, individuals are no longer surveilled just to single them out, for instance, to identify misbehavior; they are also surveilled to see how they fit in, particularly in terms of their consumption. Galič et al. (2017) summarize: "From watchwords in disciplinary societies to passwords in control societies, the point made by Deleuze in relation to surveillance is that individuals becomes [sic] less relevant as subjects of surveillance; it is no longer actual persons and their bodies that matter or that need to be subjected and disciplined, but rather the individuals' representations. It is the divided individual—consumers and their purchasing behaviour—who has become important to monitor and control" (p. 20). "Dividuals" as Deleuze (1992) called them, become surveilled, and the purpose is not just to discipline, but to "mold consumers" (p. 20). People aren't separated and isolated for who they are, they are grouped into others like them to, in a way, lose their individual identities in favor of the group. For those familiar with issues of social justice, they can see the potential problems associated with becoming the purchasing potential of their category in terms of their age, sex, race, gender, zip code, income, title, education, and any other demographics, particularly in unknown ways and sorted through machine learning or intellectual-property-protected algorithms (Bennett Moses & Chan, 2018; Carlson, 2017). Stereotyping consumers has ramifications beyond similar users buying a certain book and creeping into the way goods and services are distributed. Some get categorized as "ins" and some are "outs," which chapter 4 will describe more, through the lens of oppression.

One way surveillance theorists talk about this shift toward consumer surveillance is through a discussion of the third influential paradigm of surveillance capitalism that draws from scholarship on the information economy. The term *information economy* isn't new, with early scholarship coming out in the late 1960s and early 1970s (Porat & Rubin, 1967; Porat, 1977). The idea of the information economy is that information itself is a commodity, which has both a supply and demand side, the "supply side" being from firms and industries that gather and generate information and the "demand side" comprised of "firms, households, governments, and exports" that exchange information at a market price (Porat, 1977, p. 4). It is important to see this emphasis on the "information" economy, because it helps to see that information becomes an entity that can be bought and sold, and whether one works actively in the "primary information sector," where a business' main product is information (such as a data broker[4]), or the "secondary information sector," where a company creates information as "ancillary or 'secondary' to the production of a noninformation good" (for instance, Google, Facebook, etc.), doesn't explicitly matter. Information overall is being generated somehow, somewhere, which gains economic value (with the questions of how and where being key entry points for those interested in social justice).

Correlating more to the conversation in the book, since the turn of the twenty-first century, this "information economy" has also often overlapped with pushes toward digitalization and surveillance economies, and as this book will argue, the overlap between surveillance and the information economy has turned information workers (and often, correspondingly, technical communicators) into surveillance workers. (As I will describe in the next chapter, this can be directly those who work in positions of capital *S* surveillance and those who work more unconventionally in small *s* surveillance industries.)

Surveillance capitalism tries to capture this economy of information, particularly by corporations but also others, and explains how and why firms watch and record their users. In Shoshana Zuboff's (2019) terms:[5] "Invented at Google beginning in 2000, this new economics [of surveillance capitalism] covertly claims private human experience as free raw material for translation into behavioral data. Some data are used to improve services, but the rest are turned into computational products that predict your behavior. These predictions are traded in a new futures market, where surveillance capitalists sell certainty to businesses determined to know what we will do next." Not only does surveillance capitalism shift the emphasis of the surveillant agent away from the state and discipline, but it also transforms surveillance from a more formal state apparatus into something more commonplace, or "everyday."

This shift into something more ordinary or what can be considered "everyday surveillance" is the fourth paradigm and reflected by a widely cited definition of surveillance by David Lyon, a leading scholar in the field of surveillance studies. He defines surveillance as "any collection and processing of personal data, whether identifiable or not, for the purposes of influencing or managing those whose data has been garnered" (2001, p. 2). Important to this definition is that surveillance is *any* collection and processing of personal data (any information about someone) that is either identifiable, supposedly anonymous, or deanonymized, that is being used (or stored with intentions to use) to sort the individual into some type of category such as what type of consumer one is (e.g., Amazon recommendations), what type of financial risk one is (e.g., bank loans), or what kind of person one is (e.g., social media). Surveillance broadens to mean something that you, I, and anyone else can participate in. Surveillance becomes an everyday (as in common, normal, or routine) occurrence (Staples, 2000).

Although some such as Fuchs (2011) argue against neutral definitions like Lyon's that could theoretically allow almost everything to be considered surveillance (and he argues that it removes the focus away from power structures, which is problematic), Lyon's definition is overwhelmingly popular and also reflects a shift from Foucauldian "disciplinary" societies to Deleuzian theories of "control" societies. Instead of a disciplinary society, where surveillance likely takes place through institutional settings, in a control society, due to technological affordances, institutional control collapses and surveillance transcends and condenses in previously "private" places. As Wise (2002) states: "The barriers between home, work, school, prison, and the hospital begin to break down and run together. . . . In a control society one could conceivably be at home, telecommuting into work, taking a telecourse, be on prison leave—attached to an ankle monitoring device—and be in the hospital—attached to monitoring devices that dial in to your doctor with your current vitals—all at the same time" (p. 32). Supporting the idea of institutional implosion (or at least lowered but still monitored borders), Lyon (2007) states, "Whereas Foucault had theorized surveillance in the context of confined fixed spaces like the Panopticon, Deleuze proposed that such old sites of confinement were no longer the only or the primary sites of surveillance" (p. 60). As stated, the control society is thus about being surveilled more ubiquitously and not just by governmental institutions. Everyday surveillance transcends institutional spaces and is brought to formerly private places such as web browsers in one's own home. Surveillance becomes something not just from the state, but something that exists in the everyday, which "normal" people (typically thought of as non-"Surveillance Workers," as the next chapter will discuss) can participate. "Everyday" surveillance is a regular occurrence in postmodern life (Staples, 2000). The idea of "everyday" surveillance is crucial for this book because we move away from this top-down, governmental form of surveillance into something seemingly more unofficial or mundane. Everyday surveillance carries the assumption that not only the state can watch you, but even *you* can be surveilling someone, or watching someone else to make calculations, even if it is just to determine where to stand in a line when you walk into a grocery store.

Looking more microscopically to summarize clusters of research rather than speaking of more macroscopic paradigms, Galič et al. (2017) offer an alternative way to think about surveillance through a summary of surveillance scholarship. So much work has been done with the panoptic lens that Galič et al. (2017) described their first cluster of research as "exploring the panopticon" (p. 11). Work here includes that of Bentham and Foucault but also "offers architectural theories of surveillance where surveillance is often physical and spatial, involving centralised mechanisms of watching over subjects" (p. 9).

Galič et al. (2017) identify the second group as the "post-Panoptical theories and concepts" (p. 18), which are forms of networked, infrastructured surveillance that "involves distributed forms of watching over people, with increasing distance to the watched and often dealing with data doubles rather than physical persons" (p. 9). Theorists here are those like Deleuze and Zuboff, as mentioned, and Haggerty and Ericson's (2000) influential work on the *surveillant assemblage*, which is worth mentioning in a little more detail and almost forms a paradigm itself with its inspiration from similar research.

The surveillant assemblage is built off work from Deleuze and Guatarri's theories of assemblage and rhizomes. The surveillant assemblage reframes surveillance from more linear structures of power stemming from centralized locations such as in a panopticon to rethinking surveillance as collection of disparate and seemingly disassociated bits of data that may only temporarily coalesce to form territories and create assemblages when needed. This perspective considers the growing (and seemingly endless) gathering of information digitally from various locations for the purpose of, as described above, not just the state, but corporations. For instance, a credit check from one agency draws from multiple data points to compose and assemble a snapshot of one's fiscal responsibility at one moment in time. Like a rhizome that pops up in various (possibly unexpected) locations rather than a more linear root system stemming from one location, this data is a compilation of distributed shoots of information.

Finally, Galič et al. (2017) note entries from their third category are "contemporary conceptualisations" that branch out of earlier research (p. 26), but refine, combine, or extend the frameworks. Examples of concepts are "dataveillance, access control, social sorting, peer-to-peer surveillance and resistance" (p. 9). These theories particularly address "the datafication of society" and combine the various threads of research to join "the physical with the digital, government with corporate surveillance and top-down with self-surveillance" (p. 9). These theories try to avoid grand narratives and offer smaller-scale analysis of sites of surveillance, and include what characterizes four subsections. In the first of subcategories, in "New Forms of the Panopticon" (p. 26), scholars change the name of the panopticon to capture something different than what the panopticon represents. These types of essays were discussed earlier. The second subcategory of "Building on Deleuze: Dataveillance and Social Sorting" (p. 26), incorporates theories of dataveillance and social sorting, or the idea that our data is the thing that is being surveilled, and it is being surveilled to put us into categories of risk. As brought out earlier, we are no longer individuals, but "dividuals" subjected to data surveillance and lumped into categories of others with similar reference points. This work especially examines the processes and associated ethics of new technologically mediated sites of surveillance. The third subcategory "participation and empowerment" looks at more voluntary surveillance where people willingly choose to participate in some forms of surveillance like social media and online games (p. 29). This area is useful for thinking about the benefits of surveillance and crosses boundaries with resistance because if some forms of surveillance are voluntary, then "still individuals can, at least to some extent, resist and refuse, mainly by finding alternative ways of using technology that is increasingly accessible to him or her" (p. 29). Finally, the fourth subcategory is "Sousveillance and Other Forms of Resistance" (p. 31), which explores the way that those who are surveilled can resist, either through sousveillance (Mann, et al, 2003), which turns the eye back on the watcher, or through obfuscation or artistic expression (e.g., see Ingraham and Rowland, 2016).

A Brief Overview of Surveillance and Technical Communication

Painted with a broad brush, TC encompasses a diverse range of scholarship related to the communication of technical and technological concerns (Sullivan & Porter, 1993, p. 413). Technical communicators don't just have to work heavily with technology if their work is technical (e.g., a cookbook writer), but technology is often involved. Thus, a technical communicator in the field of TC is someone who can be versed in both the communication of technical information and the various platforms that can be utilized to deliver this information. For instance, a technical writer making a cookbook may be more concerned about communicating the content of technical information such as a recipe on a printed page (arguably still technological), but a web designer may be focused on communicating information through the mediated, technology platform. Both could be technical communicators.

Technical communication as a field isn't limited to the work of a technical communicator though. The field is also a robust place for scholarship exploring the applied and social impact of TC work, as well as a place for praxis and teaching and training upcoming communicators. In just three examples, TC explores and supports structuring rhetorical arguments and deliberation such as in an environmental context (Coppola & Karis, 2000), as a problem-solving activity for the workplace and social contexts (Johnson-Eilola & Selber, 2013), or to promote student awareness of intercultural communication (Thatcher et al., 2011). These examples are in no way comprehensive.

More broadly, Kimball (2017a) emphasizes that, TC is not just a field or job category, it is also an activity. He states, "[T]echnical communication is not just a profession, but *an activity that manages technological action through communication technologies, including writing itself, in a particular setting and for particular purposes*" (p. 346) (emphasis in original). In this added layer, being a technical communicator is more than just an official title or field; it is an activity that manages information and technology through technology and involves knowing the rhetorical situation to deliver effective messages. This definition is less formal, and for instance, those who create helpful videos on YouTube can be considered technical communicators. This book is situated closer to Kimball's work.

It is noteworthy, and a disclaimer, that this book also encourages a liberal definition of the activities that technical communicators can be engaged in to think about ways that also approach technology users as technical communicators. Given that contemporary technologies are ubiquitous, and if assumptions about what it means to be a technical communicator are broad, we can potentially see a lot of technical communication work by people who are not always considered technical communicators. This topic will be explored further in chapter 2, but for TC, this encourages reflection on how one defines and defends the boundaries of what can be TC work, and it also invites those from outside the TC field to come in and see how one could benefit from practical, humanist, and social work being done in TC.

Returning to the connections between TC and surveillance, surveillance isn't a new subject for scholars close to TC, and there are many scholars writing about the intersection of surveillance, communication, and technology. In particular to this book, TC scholarship includes robust considerations of surveillance, which can be categorized into at least eleven areas of emphasis.

- First and second, surveillance often pops up in institutional conversations about (1) the workplace or (2) schools (Amidon & Blythe, 2008; Banville & Sugg, 2021; Beck & Hutchinson Campos, 2020; Clark et al., 2012; Dautermann, 2005; Duncan & Hill, 2014; Fairweather, 1999; Machili et al., 2019; Pigg, 2014; Spinuzzi, 2007; Walsh, 2010; and Zhang et al., 2020).

While two different institutions, these spaces often overlap because frequently a mix of both are discussed, such as Dautermann's (2005) discussion of monitors in a classroom in China that borders on both a professor's workplace but also school.

- Third, state/government surveillance has also been of note, also with varying topics of focus. Government surveillance is a large piece of the surveillant machine (Weller, 2012), so it is not a surprise that this area is also well-represented in TC scholarship. There are various mentions of the government trying to keep track of people in immigration contexts (Bivens et al., 2019; Sánchez, 2020) and through the census (Li, 2020), as well as surveillance for risks to safety like terrorism (Gibbons, 2018; Wilson & Wolford, 2017). How government surveillance is portrayed in reports has also been reviewed (Markel, 2009b), and others have critiqued policies that involve surveillance (McKee, 2011) and addressed threats to research due to surveillance (McKee, & Porter, 2010).

- Medical surveillance is a fourth category. Some topics range from the more procedural discussions of surveillance activities in the health-care field such as postmarket surveillance with pharmaceuticals (Bonk, 1998; Kessler & Graham, 2018), health tracing (Teston, 2012); genealogy databases (Woods, 2021); customized surveillance against cancer (Turner, 2005); or the state's surveillance of health (Ding 2009) (which also blends into government surveillance). There are also critical discussions of other surveillance in health-related areas such as that health researchers need to be careful about ethics and their role in surveillance (De Hertogh, 2018; Opel, 2017), and that health cards can become instruments of surveillance (Coogan, 2002).

- Fifth, surveillance has also been approached through the gaze at women's bodies through topics like breastfeeding (Hausman, 2000), ultrasounds (Frost, 2020; Frost, & Haas, 2017), or fertility and period-tracking applications (Novotny & Hutchinson, 2019). These areas can also overlap with medical surveillance, but there is a strong connection to women's bodies as well.

- Sixth, there is also a cluster of research that discusses surveillance by looking through specific technology examples. Some of these overlap with other areas such as school and course/learning management systems (Schneider, 2005; Duin & Tham, 2020) or the workplace, such as interface design facilitating workplace surveillance (Mirel et al., 2008), but still have a good portion focused on the technologies that enable surveillance practices. Still other research looks at how increased connection brings more surveillance (Verhulsdonck et al., 2019) or how empty state pages encourage self-surveillance (Gallagher & Holmes, 2019). Frith (2019) looks at RFID; Tham et al. (2020), Tham and Hill (2020), and Hutchinson and Novotny (2018) either mention or focus on wearables, and deWinter and Vie (2016) look at power, surveillance, and games.

- Seventh, another focus that has touched on surveillance is the study of surveillance as part of pedagogy or research such as students looking at surveillance of gamers (Haas, 2012), students examining legislation around surveillance technologies (Agboka, 2020), and the ethics of surveillance as a class topic (Whalen, 2018). Beck (2013) also brings out that surveillance should be a concern in the future of scholarship going forward. The Digital Rhetorical Privacy Collective (n.d.) also aggregates surveillance pedagogy examples. One short but interesting thread was a debate on teaching and as to whether surveillance could or should be shortened to the word *surveil* (Bell, 1994; Bush, 1994).

- Eighth, some briefly mention that surveillance is involved in the work of those under study in a researcher's work such as Boeing's work with "Maui space surveillance system" (Richards & David, 2005) or planning done at the Research, Development, Test, and Evaluation Division (NRaD) of the Naval Command, Control and Ocean Surveillance Center (Cathcart, 1997). Surveillance isn't the primary focus of these articles, and the project may be more ostensibly environmentally related like space and oceans, but their presence in TC literature shows how technical communicators can be involved in surveillance activities that can still fall into forms of surveillance.

- Further, ninth, work that mentions that Michel de Certeau's (1984) theories of tactics (as will be discussed in greater depth, later in the book) are formed in the cracks of surveillance have also elicited short conversations about systems of watching in some scholarship (Colton et al., 2017; Edenfield et al., 2019; Petersen, 2019; Sherlock, 2009).

- Tenth, there are those who focus on consequences of surveillance. Zhang and Saari Kitalong (2015) talk about surveillance's influence on creativity, and Zachry (2008) talks about the consequences of classification, which can assist in surveillance. That isn't to say that other scholars don't also address consequences of surveillance, too, but consequences are a strong core of this category. In adjacent scholarship, Beck's (2015) "invisible digital identity" borders on TC and describes the bodies of data that accumulate.

- Finally, eleventh, there are those who draw on the relationship between Foucault, panopticism, and/or disciplinary power. For instance, there is talk about panoptic surveillance design in the slavery in Caribbean plantations (Ramey, 2014), the panopticon and visual design (Barton & Barton, 2016), Foucault's relevant conversations about Victorian-era surveillance of sexuality for workplace harassment policies (Ranney, 2000), and how cookbooks serve as a form of discipline (Moeller & Frost, 2016).

There are three side notes to this conversation. First, it is worth mentioning that computer and writing scholarship, with adjacent and often overlapping TC scholarship, also addresses surveillance (Anderson, 2006; Beck, 2015; 2016b; Burley, 1998; Colton, 2016; Crow, 2013; Day, 2000; DeVoss et al., 2005; Fielding, 2016; Gonzales & DeVoss, 2016; Hawisher and Selfe, 1991; Hawkes, 2007; Healy, 1995; Janangelo, 1991; Kimball, 2005; Marsh, 2004; McKee, 2011; 2016; Moran, 1995; Purdy, 2009; Reilly, 2016; Reyman, 2013; Tulley, 2013; Vie, 2008; Vie & de Winter, 2016; Walls, 2015; Young, 2020; Zwagerman, 2008). This is particularly important to note, because Foucault (1977) panopticism frequently shows up in this research, too, and helps illustrate the frequency of using Foucault's work. For instance, in the 1990s, Hawisher and Selfe (1991) identified the panopticon as relevant to explaining electronic forums, and they remind the audience that work done through a network can also be used to keep track of students.

Janangelo (1991) discusses how in the online classroom, watching can flow not just from teacher to student but also from teacher to teacher and student to student. Day (2000) also illustrates how faculty members are watched by administration in a panopticon-like way. Healy (1995) similarly draws on the panopticon and demonstrates that writing centers with an online component can monitor both the tutors and the students working in these spaces. Healy concludes that with too much monitoring, "Panoptic principles" (p. 190) can threaten tutors' desires to take risks, try new strategies, and pursue intuition. More recently, Zwagerman (2008) and Beck (2016b) use the panopticon as a lens for plagiarism detection software; McKee (2011) uses a variation of the idea of the panopticon to show how internet activities can be monitored by unknown government or corporate actors, and Beck et al. (2015) discuss panoptic power in course/learning management systems. Even when scholars move away from Foucault, they sometimes reference the metaphor to help ground newer principles in foundational work, like when Beck et al. (2015) discuss how the panopticon doesn't fully account for how bodies are abstracted as data doubles and Crow (2013) focuses on how surveillance paradigms are moving away from the panopticon.

The second side note is that privacy is also a large subject of concern for technical communicators. As will be explained, this is complementary to surveillance but places emphasis on privacy and not surveillance. Due to the constraints of this book, surveillance research remains the focus of this book.

Finally, a third point is that despite the presence of the word *surveillance* in many of these pieces and the ability to identify emerging categories, many of these works only mention the actual word *surveillance* once, and if the word is used, there is often (but not always) less interrogation of the meanings and dimensions of surveillance. While it is useful for a researcher to look back at the collection of works to see what procedures scholars in TC have labeled *surveillance*, it is also useful to work with surveillance's complex dimensions and increase vocabulary and discourse about what it means to engage in surveillance practices in contemporary times. This is especially the case because oftentimes, just as in the case of Edward Snowden, it is a technical communicator who is engaged in the practice of surveillance, and a technical communicator, whether they know it or not, can be engaging in the practice of watching or taking notes on someone else. This is consequential because information used to manage others is the foundation of contemporary surveillance practices (Lyon, 2001).

Where Technical Communication and Surveillance Overlap

All these different areas might seem overwhelming when introduced together in such a brief and compact space, but there are some definite overlaps between surveillance and TC. It is useful then to explore some more macrolevel connections to provide a foundation for concepts teased out in the rest of the book. Based on the scholarship presented in this book, surveillance and TC are particularly suited to each other for two complimentary reasons: (1) the expertise of technical communicators and (2) the scholarship of technical communicators, including where there is room for future connections. Both spaces are reflected in the review below.

The Expertise of Technical Communicators

Technical communicators have competency and expertise that puts them squarely in the realm of potential to do surveillance work, especially due to their role as (1) information workers and (2) their technology expertise. To explain both, first, in an information economy (as we briefly just described), technical communicators are often information workers and managers of digital data. Porat (1977) describes "information workers" as those "essentially paid to create knowledge, communicate ideas, process information in one way or another transform symbols from one form to another" and whose jobs are primarily to manipulate information for either the purpose of intensive mental work creating new knowledge or the more routine tasks like inputting information into a computer (p. 3). So, too, are technical communicators who communicate (technical) knowledge either about or through those technologies (as described about technical communicators by Sullivan & Porter, 1993, p. 413).

Correspondingly, as the information economy grows, this demand can lead to the increasing development of technologies built to gather information. Thus, there will also be an increasing demand for technology and technical workers that can produce, record, manage, and store that data. As in the idea of everyday surveillance, any gathering of information for management purposes can be considered surveillance, so simply working with someone's information could be considered a process of surveillance. Due to the technical expertise of technical communicators, then, technical communicators working in information may find themselves working in surveillance-related jobs, whether they meant to or not. Contemporary information communication technologies are even said to be synonymous with surveillance (Sewell & Barker, 2007).

It is especially possible that a technical communicator may not initially intend to gather information for what could be considered surveillance purposes, due to the creep of surveillance practices. According to Andrejevic and Gates (2014), "function creep" refers to the "[c]ontinuous repurposing of information initially gathered for other purposes," made easier through "storage, sharing, and processing" of increasingly digitized information (p. 189), and is a slow way surveillance is often introduced. For instance, some technologies, such as stingray devices, might seem readily surveillance-enabling. Stingrays can act like cell phone towers and can intercept calls from nearby mobile phones, which can in turn collect communications and metadata from whoever connected to the technology. Popular in law enforcement contexts (see American Civil Liberties Union, 2020a), these technologies work directly in surveillance markets. Other technologies, though, have less of a surveillant edge. For instance, the acquisition of health information by a Fitbit could still be murky when the contextual integrity[6] of the data's privacy is changed and the fitness device's records are moved or sold to third parties. It was clear that the device was supposed to be monitoring someone, but the purpose of the use of data may shift from monitoring health to marketing products. The function has crept.

The Thematic Concerns of TC Scholarship

TC is also a solid foundation for surveillance because TC scholarship is already complementary to surveillance in at least eight different ways.

Contextual and Situated

First, surveillance and TC are already complementary due to the way both groups emphasize the importance of context and situation. Both groups argue that to understand any phenomenon, you must also know the larger circumstances around it. In the language of surveillance scholars, these claims largely hinge on the ideas that surveillance differentially impacts people, and surveillance must be evaluated through particular constellations of circumstances. Drawing from ideas of contextual integrity for privacy (Nissenbaum, 2009), surveillance is also contextual because appropriate and norms for watching and recording others is not only a moving target but also highly contested between stakeholders. Police profiling targets can be justified by one party but opposed by others. Watching someone in reaction to an imminent threat is different from wide swaths of indiscriminate watching, especially those that target certain communities. ("Imminent threat" is also open to interpretation.) Spies, lurkers, corporations, and friends can all be said to watch each other, but sometimes the surveillance is welcome, and other times it is detrimental.[7] Just the terms these entities are referred to, as well as the term surveillance itself, are loaded with semantic baggage.

To understand surveillance means to examine it on a case-by-case basis for an ethical and just assessment, and a useful exploration of surveillance is one that considers the "moving parts" so to speak, that exist in any given community. It is nearly impossible to make universal statements about surveillance (beyond that it is not experienced the same way by different people). Sometimes surveillance is good, and sometimes it is bad; who it helps and who it hurts is open to interpretation, and it can even help and hurt at the same time. What can be good for some is not good for others, and further, even "acceptable" surveillance isn't static and does not stay "acceptable" at all times.

TC scholars are poised to jump into a conversation about surveillance because they have already conceived of their own work as contextual, or "situated." Communication scholars like those in TC regularly assert that we must look at different situations to understand and interpret phenomenon, as well as recognize our own positionality or ethos, and, as rhetoricians, communications scholars are always thinking of audiences and talk about the constant preoccupation of the situations, conversations, or "parlors" one walks in and out of (Burke, 1941), or the rhetorical context that surrounds any one instance of communication. The concept of being "situated" implies a "situation," and definitionally a situation is just an ephemeral assemblage of circumstances one finds oneself in. For each situation, location matters (Rich, 1984), and while long-standing theories and approaches are useful, one situation is likely different from another, and situations are best recognized for their uniqueness in context (Ede, 2004) and one's own positionality (Walton et al., 2019). There can be best practices for any type of correspondence, but as in the case of writing a business report, how a document is written and interpreted is pragmatic and essentially depends on specific organizational contexts (Kalmbach, 2007). Similarly, words like *rhetorical context* or the *rhetorical situation* (Bitzer, 1968) also reflect the idea that we must understand the time, space, place, and convergence of speaker, message, and appeals to make sense of any particular utterance.

Surveillance's *Rhetoric*

Second, then building on this talk about rhetoric and how surveillance depends on context, I would also add that surveillance itself is an argument in itself, a rhetoric invoked sometimes and not at other times, and an argument needing assessment. As experts on lines of reasoning and evoking the right appeals at the right place at the right time, TC scholars make excellent candidates not only for analyzing and critiquing the argument, or rhetoric, of surveillance itself, but also for critiquing the deployment and fitness of that rhetoric.

To explain, rhetoric can entail many things, from an argument, to a speech or written word, to the study of a speech or the written word, to the use of language to inform or persuade, to the persuasive effects of that language, to the relationship between language and the creation of knowledge, to classification of speech types, or of empty jargon (Bizzell & Herzberg, 1990, p. 1). Aristotle, a key figure in the narrative of ancient rhetoric, argued the popular refrain that rhetoric is "the faculty of observing in any given case the available means of persuasion" (Aristotle, 1990, p. 153). While the level of emphasis on the term *persuasion* is in flux with context to the degree that one is trying to "persuade," *rhetoric* is sometimes simplified to mean attention to the ways a speaker (broadly interpreted) delivers a message (broadly interpreted) to an audience (primary, secondary, tertiary, etc.). But we can talk about rhetoric in the classical sense, rich with history, ancient philosophers like Aristotle, and terms and classifications, and we can talk about its genesis of these concepts through time to its more modern form where it looks like a pastiche of both the ancient and contemporary.

Rhetoric for this conversation draws on the work of twentieth-century rhetorical theorists who emphasized that it is made of arguments of persuasion that can be used as a tool to flesh out ideologies and assumptions of rationality (p. 900). Rhetoric can even be viewed as epistemic (Scott, 1967), with its assertions of truth. Rhetoricians and their rhetorics then, full of ideologies in their discourses, employ and deploy certain strategies and symbols to certain audiences, in certain situations, in attempts to affirm certain truths, incite particular actions, reactions, and identifications to reach particular goals (Burke, 1945; Perelman, 1973). For use in this conversation, a rhetoric can be understood as collection of multiple rhetorics, full of strategies (rhetorical appeals, language, symbols, etc.) that assemble to create a strategic choice. As previously noted, arguments exist in context, and an argument is really but one assemblage of rhetorics among other possibilities, that is deployed and accepted or resisted as some form of truth by the audience or participants.

Through this lens, surveillance and rhetoric are intricately linked, particularly in that surveillance is a rhetoric, or in other words, an argument in itself, imbued with strategies and symbols to accomplish a goal. Oftentimes (although not always[8]) this goal can be the epistemic pursuit of "truth" and "reality," in which surveillance, or the watching over and recording of information provides some type of clarity of what something "truly" is. Finn (2012) illustrates the rhetoric of surveillance through his discussion of reality shows, and when it comes to watching others on TV, he argues that surveillance becomes the way the narrative is legitimized. Finn comments, "[S]urveillance serves as a rhetorical device to guarantee the veracity of the events depicted" (p. 75). In the case of reality shows, surveillance, or the watching and the visibility of the contestants, is the answer to how we find out "the truth." We not only watch what the participants do during a more formal challenge, but we also see them in their "everyday" interactions to see what they say about themselves and others and how they interact.

Although Finn does not comment on the impact of a producer's editing of a show to co-construct "truth," he does argue that this watching does bring in "a particular authoritative weight to the surveillance camera, where its seemingly automated, anonymous and omnipresent gaze functions as a harbinger of truth" (p. 76). Surveillance thus becomes the way the framing of a particular participant is legitimized, and in the reverse, surveillance is legitimized by the assumptions that viewing someone else finds the "truth." Extending beyond reality shows, Finn also notes that surveillance rhetoric also allows for claims of truth in other situations with closed-circuit television (CCTV) cameras (such as governmental security) or tabloids that aim to show celebrities and high-profile workers in their "normal" settings. We watch others to know who they "really" are.

Zooming out from Finn's argument to the context and theme of this book, surveillance isn't just a rhetoric of truth-making for reality shows, CCTV footage, and tabloids. Surveillance in general is a rhetoric, or one argument, that shapes actions and reactions when dealing with a particular situation. If truth comes from the watching and evaluation of others through surveillance, then surveillance becomes an acceptable practice. If a corporation wants to argue that a demographic "truth" will be revealed through data analysis, and that argument is accepted by those in the situation, then the acceptance of the practice also legitimizes the process of watching and monitoring. What we begin to see repetitively also shapes our attitudes toward surveillance in general, and we begin to see surveillance as acceptable, with surveillance contributing to our "understandings about certain aspects of visibility in daily life, and in social relationships, expectations, and normative commitments" and what actions we accept as more and ethical (Lyon, 2017, p. 829).

Practices of surveillance are further carried out through symbols, particularly visual ones, or what Finn (2012) might describe as a surveillance aesthetic. In his work, Finn explores how searching stock footage archives using the word *surveillance* illustrates a range of symbols that communicate surveillance practices with a heavy emphasis on CCTV cameras. A search of surveillance is simultaneously a look at all types of surveillance cameras in operation. A look at "surveillance" through other stock footage archives like Adobe Stock (n.d.) also reflect this similar attention to CCTV cameras but other types of machine vision like idealized dashboards of digital databases, facial scans, drones, and doorbell cameras. We also see more physical representations of watching such as law enforcement workers in control rooms, shadowy agents lurking at a distance, visuals of law enforcement officers with sunglasses, or the picture of the dark, hoodied hacker with a slightly more threatening aesthetic (Gallucci, 2017). These symbols of surveillance not only show up as a stock footage surveillance aesthetic, but they proliferate in any type of image search on any web browser. These images become symbolic of surveillance ideologies and tell us how we react to it. We see how it is carried out, who is doing the surveillance, and how it is assimilated into our lives. The symbols of surveillance are key to its rhetoric, and Finn argues that these images, especially with these stock footages, "position surveillance as a banal, commercial concept: it exists as a generic category to be used in the construction of visual content . . . [Images] are de-politicized, de-historicized, and convey nothing of the problems and tensions associated with the practice of surveillance" (p. 74). The hacker as surveiller gets demonized through their shadowy depictions, but other types of surveillance are normalized as part of the day-to-day life experience. Yes, police conduct surveillance, but this is to "find the bad guy." I would also argue though, that surveillance is an everyday activity, so that is why a critical eye is needed—because it is in fact, pedestrian and often just part of a day's work.

Key to the meaning of this conversation though, is that surveillance—its symbols and its presentation of a solution—is just one of many rhetorics that could be deployed. Surveillance, or the watching of others, can be just one way to find "truths," or just one way to reach goals. Surveillance as a normalized, banal, commercial concept is just one reaction that the audience can have. It is up to critical thinkers like technical communicators to identify and critique consequences of surveillance. For surveillance, this is particularly important because the assumptions that watching others either directly or at a distance through technology to arrive at the "truth" is both negotiated and also not an uncritical biproduct of surveillance. In the case of the watching of police, an officer may find surveillance footage of a robbery suspect so that some sort of "truth" comes to light. But in the case of crime statistics and the watching of historically marginalized communities more frequently than other communities, "truth" can also become a distortion of spending more time looking in one place for certain crimes rather than focusing on others. Or if an employer is watching every standing, nonscanning moment of a warehouse employee or every keystroke of a remote worker to determine time off task, the "truth" about productivity may be only half constructed, without accounting for a hunt for a missing item or an extended gesture of customer service.

Recognizing surveillance is an argument, or a rhetoric, then, becomes central in a critique of its use. Thinking of surveillance as a choice opens up questions like, why should we use surveillance as tool for establishing some type of "truth?" What are the alternatives, if any? If there are no alternatives, what are the benefits, consequences, or harms, and how can these be maximized or minimized? While these questions are very difficult to come to in isolation without context, it is important to first recognize what surveillance is and then rhetoricians/technical communicators can work through evaluating situations to see the consequences of the rhetoric's deployment.

Surveillance Harms

Third, surveillance and TC are complementary to each other due to what TC and surveillance might consider "harms." Pinning down surveillance harms can be challenging because different people can view and experience the same situation in various ways. That is certainly problematic because if everyone viewed a particular instance of surveillance as bad, then there could be enough momentum to end that practice. However, if, for example, some view drone use as good, and some view it as bad, then the potentially harmful surveillance practices have room to proliferate, since shared goals are useful for making systemwide changes (Green, 2016). A big problem with surveillance then, is that what is "problematic" is not always problematic for everyone. To see one's own positionality and that of another requires a thoughtful consideration of what is being done and who it effects and how. Putting these complexities in the background, we can still forefront surveillance harms in four categorical ways: harms to social justice, civil liberties, human rights, and chilling effects.

For harms to social justice, surveillance can cause the unequal distribution of goods and services, target historically disadvantaged communities, or create unequal economic pay structures where the marginalized end up with higher payments (Weinberg, 2017). More about social justice and harm is explored in chapter 4.

Surveillance can also be consequential for matters of civil liberties and human rights. There is a strong thread of surveillance research that questions the limits of surveillance and the affordances that individuals and groups should have against it. In simple terms, civil liberties are the protections afforded to individuals by documents like the US Constitution against actions of the state or governments, and human rights are motivated by the assumptions that there are affordances that any person should be given, by virtue of being alive. Snowden drew on areas of this rhetoric to justify his own disclosures of government surveillance. In his biography, Snowden (2019a) often talked of civil liberties and claimed he grew uncomfortable with the shift from targeted surveillance (such as the focused monitoring of suspects) to mass surveillance of entire populations not under immediate suspicion. This sort of surveillance allowed the government to "collect all the world's digital communication, store them for ages, and search through them at will" (p. 1). Snowden asserted that he felt these actions were not only "in gross contravention" of the Constitution but also the "basic values of any free society" (p. 3). Snowden thus argued by way of the Constitution for America or through any law for a country that professes to be "free" that it was the state's obligation that they protect certain fundamental privacies and not permit excessive surveillance.

Koops et al. (2017) illustrates further that beyond civil liberties, privacy and/or an absence of surveillance have been linked to human rights and essential right to keep certain aspects of one's life private. Privacy then is not just an affordance upheld by a state, democratic or otherwise, but it is framed as something that every human should have. The acts of surveillance degrade affordances of privacy and rights of world citizens not to be unduly under watch of their governments.

"Chilling effects" can also be a harm (Richards, 2013). Chilling effects result when people worry or are discouraged about exercising their legal rights, which may encourage behavioral changes. While behavioral changes may not be these "bodies in the street," surveillance, or a fear of surveillance can alter one's lifestyle in hard to quantify but still detrimental ways. Chilling effects are particularly useful for making the point that just because surveillance might not have a visible harm (such as a body in the street), that doesn't mean that surveillance can't be harmful or problematic. Just because there might not be the threat of physical harm such as what might exist if an abusive spouse threatened their partner with a gun, "chilling effects" might cause the same ex-spouse to avoid going out, avoid going online, or avoid talking to specific people because they are afraid their ex may be watching.

Social justice, civil liberties, human rights, and chilling effects in particular are matters that technical communicators can address due to the field's focus on reacting to harms in general through a social justice lens.[9] Jones and Walton (Jones, 2016) explain that social justice research in TC "investigates how communication broadly defined can amplify the agency of oppressed people—those who are materially, socially, politically, and/or economically under-resourced" (p. 347). This can be done by addressing position, privilege, and power (Walton et al., 2019, p. 7), and "the oppressed can be from any demographic, ethnic group, age group, religious background, or culture" (Jones, 2016, p. 347). The field has taken this focus as an important goal and has made a concentrated effort to address structural and systemic oppression. As summarized by Colton and Holmes (2016), "Social justice research currently strives to recognize injustices within institutional contexts in order to call for the revision or reimagination of these contexts" (p. 5). Social justice approaches are particularly useful for assessing social concerns, surveillance's chilling effects, as well as human rights and civil liberties violations, and chapters 4 and 5 further explore the connection between the approaches and the assessments.

Cultures, Targets, and Agents

Surveillance is also thus cultural because it is contextual and situated, and surveillance isn't carried out or felt in the same way in every community. By cultural, I mean that, as stated, surveillance must be evaluated in the various contexts or cultures where it is executed, interpreted, and consequential. These variations are key to the area of cultural surveillance studies. Monahan (2011) notes that cultural lenses help us see that "experiences of surveillance differ by population, purpose, and setting" (p. 497).

For those *targeted* by surveillance, it is particularly important to see how one is affected by the situation of visibility. It is worth knowing how one is being managed with their body and information, especially because surveillance practices do not "simply slow down or single out people considered risky—it also accelerates and augments the experiences of people considered to be of commercial value and low risk" (p. 497), and thus surveillance amplifies social inequalities. Corresponding to social justice then, surveillance draws attention to an examination of those under the critical eye of surveillance to see who is targeted and how they are affected. We need to see who is targeted and why in the context of particular communities and cultures where the surveillance sites are situated.

Attention to targets is painfully inadequate, however, without a parallel attention to the agents that conduct surveillance. Those who carry out surveillance are integral to any cultural and contextual assessment of surveillance. Monahan (2011) notes that cultural surveillance studies not only look at "those subjected to it," but also at "those appropriating [surveillance] for their own purposes" (p. 497). As noted earlier, agents of surveillance can come from a variety of places, be it the state or the "big brother," an institution like a school or hospital, a corporation and the "little brothers" (McGrath, 2012, p. 83), a friend, or the technical communicator; there can be extensive lists of agents that, again, go back to the context of any given situation where surveillance principles of visibility, sorting, and control can be identified. (This book makes it a point to reiterate how an everyday worker or even oneself can be that agent.)

To assess the targets and critique the agents, Dubrofsky and Magnet (2015) note several approaches to extracting the cultural dimensions of surveillance in their collection, *Feminist Surveillance Studies*. The authors note that the essays they feature draw from "feminist theory, critical race theory, critical cultural studies, communication theory, media studies, critical criminology, and critical legal studies" (p. 2) to interrogate the various ways surveillance targets particular communities.

Dubrofsky and Magnet's frames pair nicely with work being done along similar lines in TC. In their work on threads of TC scholarship, Jones et al. (2016) provide a useful (but not exhaustive) list of antenarratives that guides other culturally focused work, including such themes as feminism and gender studies, race and ethnicity, international/intercultural professional communication, community and public engagement, user advocacy, disability and accessibility, many of which overlap cultural studies of surveillance. These antenarratives could be used to examine spaces of surveillance or cross-cutting themes of surveillance if communicators examined how regimes of watching impact everyone under watch. Looking at how surveillance (disproportionally) affects various communities is key to the power of using surveillance as a lens of analysis.

One more point about this more cultural framing pertains to cultural representation and how surveillance is imagined in a more theatrical sense of the word *culture*. The "surveillant imagination" is what plays out in movies, books, plays, television screens, and YouTube videos. It is how we *imagine* surveillance. Snowden (2019a) even illustrates the power of cultural imagination himself when he talks about the physical location on the first day of his job at the "Old Headquarters Building" (OHB) of the CIA in McLean, Virginia. He commented that the OHB houses more workers than the "New Headquarters Building" (NHB) in Langley, Virginia, which is typically shown in movies (p. 121), but the OHB is "far less ready" for its Hollywood close-up than the marble floors of the NHB. Snowden described his office as being "a grimy, cinder-block-walled room with all the charm of a nuclear fallout shelter and the acrid smell of government bleach." In my own experience, surveillance work that Hollywood might transform into something exciting and glamorous takes place in an average office with spreadsheets on old computers under fluorescent lights in windowless basements. This example also goes back to surveillance rhetoric and the surveillance aesthetic that circulates ideas through symbols and helps shape attitudes toward surveillance.

This imagination is very consequential, though, for at least two reasons: the tropes presented limit other imaginations, and in the imagination, consequences tend to be imagined very severely. To illustrate, one of the most referenced depictions of surveillance comes from George Orwell's novel, *1984*, where the leader of Oceana, Big Brother, constantly "is watching you" (Orwell, 1950, p. 5). No matter where citizens go, the state watches them, waiting for evidence of a misdeed. But, just as the panopticon has encouraged scholars to think in one paradigm and not the other, works like *1984* also encourage thinking about surveillance, in state-sponsored ways, looming and brooding over its seemingly helpless citizens (Kammerer, 2012). Even more contemporary, shows like *Black Mirror* also skew toward a dystopian view of surveillance (Elnahla, 2020). Both of these visions start funneling ideas of surveillance into more negative directions. They turn attention away from possible positive benefits of surveillance, which limits the imagination and encourages us to think in negative terms about surveillance. This may be a good thing if surveillance is oppressive, but it can also be detrimental not to see the conflicting good and bad of surveillance. Again, surveillance can proliferate because it also has benefits.

These directions lead to the second consequence of cultural depictions of surveillance: the severity consequences. This isn't just a matter of skewing bad, but also skewing *really* bad. Big Brother, dystopian technologies, and surveillance become the enemy. While those against surveillance may see these depictions as a step forward, they are also not wholly realistic. As Wise (2016) notes, surveillance is often less cinematic than it is portrayed on-screen. He states, "[A] brutal centralized state regime degrading and abusing its population" is not only cinematic but easy to rally against; however, tendencies of surveillance capitalism and "[a] proliferation of state and private institutions instilling discipline by gathering information, rendering judgements, and encouraging the production of normative subjects" is "less clear, less cinematic" (p. 78). But surveillance is often less catastrophic than that and more "everyday," without a body count. Contrasting this with the conversation about the neutrality of the surveillance aesthetic in stock footage, one conclusion that can be drawn is that cultural depictions of surveillance overall don't represent the complicated nature of surveillance, which is again, where TC scholars as cultural theorists fit well in engaging in the larger conversation. If we look for surveillance only in its deep dark places, we may also miss its bright parts, which encourages its proliferation.

ETHICS FOR EVALUATION

The fourth, connection is the complementary relationship between surveillance, TC, and ethics. In surveillance research, as described by Eric

Stoddart (2012), ethics of surveillance are often split into two categories, either: (1) a rights-based approach or (2) a discursive approach. Rights-based approaches position ethics as a matter of fundamental rights that should be granted by the state. Arguments of data protection and privacy fall into this rights-driven category, with data protection carrying the belief that "a balancing is required of the right to privacy, economic interests and individuals' well-being" (p. 370), and with privacy drawing on the belief that individuals should be able to live a life without an unwanted gaze.

The discursive approaches contrast the beliefs of rights with the assumptions of what is the best course of action in a situation—ethics then aren't definitive rights but a continuous exercise. Stoddart continues, "Ethics are not the outcome but the process itself" (p. 373), and he illustrates this by referring to public health and the identification of "at-risk" individuals through surveillance or workplace surveillance in different rhetorical situations.

Marx (1998) also offers another entry point into surveillance and ethics when he provides a table of questions one can use to determine the ethics of surveillance. His twenty-nine questions range from an interrogation of the means involved, to the data collection context, and the uses of that information. For Marx, it is important to assess a range of considerations such as if surveillance procedures cause any harm or cross any boundaries, and if the information is obtained with consent and carefully maintained. Emphasizing an analysis of conditions like agent, act, site, target, motivation, information, paradigm of surveillance, and consequences at any given time also emphasizes the situated and discursive nature of surveillance.

Similar to surveillance, TC scholarship can also be discursive and is especially poised to enter the conversation of ethics due to its historic focus on ethics and the steps that a technical communicator can engage in to make ethical decisions. For this book, ethics will also be used for a system of analysis to start breaking apart the contradictory nature of surveillance. While social justice scholarship may be better poised to answer how to react to surveillance (e.g., coalitional action, as will be discussed), ethical evaluations are especially helpful for evaluating the duality of surveillance. Surveillance can obviously be harmful, as when it's disproportionally aimed at particular communities. But surveillance can also be helpful, and it can help reach goals. Monahan (2011) concludes that surveillance can "serve democratic or empowering ends if it brings about openness, transparency, accountability, participation, and power equalization among social groups and institutions" (p. 498). Surveillance can also help purchasing, health and elder care, utility monitoring and sustainability, or inventory control. But again, even these same "goods" also turn problematic when data is misused, networks get compromised, or behaviors are tracked. Surveillance can be as helpful as it is harmful, and even both at the same time. Ethics will also be discussed more in chapter 4.

Effective and Affective

A sixth complementary node between surveillance and TC is surveillance's potential to effect and affect its audience. Surveillance not only asks for certain actions, but it also encourages certain reactions as well. To first explain effect, as I have argued, surveillance is a rhetoric, or an argument. Previously, I focused on how surveillance is an argument of how we can come to know "truth," but it is also a rhetoric of action, and its argument tells others how to act and react. Surveillance does not just tell us the "truth," surveillance also tells us what to do with that "truth." It asks for an effect. If a data broker sells information to a magazine company and has recommended that a zip code and age make a person more susceptible to buying a particular product, then this "fact" may motivate a publisher to send out a free copy of *Garden & Gun*, whether or not the person would have any interest in either gardens, guns, or the culture surrounding the magazine. The surveillance data asked for a particular action based on the assumptions of its algorithmic parameters. Overall, the whole motivation for surveillance capitalism is built on hopes for an effect and that gathering more data about someone or someone like them will help predict future possibilities such as purchases. Data brokers sell data to companies which supposedly represent types of consumers that can be used to market products. Surveillance "truths" in general, then, are summaries of data gathered for some type of action, or effect.

Closely related here, surveillance is not just something that might cause an effect, it is something that creates an affect, particularly because it can cause people to be moved emotionally. As felt experiences, people often physically feel very close to their data, and disruptions to privacy from surveillance can cause fear, insecurity, or anxiety (Stark, 2016). Surveillance can cause fear or shame more directly, such as experiences of being detained at border checks (Selfe & Selfe, 1994) or when someone is being stalked by a former romantic partner. An affect can come just by the fear that someone is watching. One particularly useful illustration of affect was when a guest speaker in one of my own privacy and surveillance courses asked everyone to open their phones and pass those unlocked and open phones to their neighbors. Instantly students expressed anxiousness and nervously chuckled about how uncomfortable and panicked they physically felt thinking a neighbor would have access to their unlocked phone. Students here "felt" privacy and/or surveillance and ultimately did not have to pass the device if they felt too uncomfortable.

In addition, as Ellis et al. (2013) argue, surveillance often goes unnoticed—"experienced on the margins of consciousness" and instead of being articulated as a direct response or acceptance of surveillance, it can be felt and exhibited, not necessarily in resistance, but also in "disruption, disfluency, and hesitation in the text of speech acts rather than clear representation" (p. 716). Richards's (2013) discussion about the chilling effects of creativity can relate to this more subtle affective response. Going back to the earlier conversation about civil liberties, Richards argues that surveillance can affect the way people act because surveillance can chill the exercise of our civil liberties, particularly around theories of First Amendment provisions for intellectual freedom. We may be granted the freedom to think and assert new ideas, but if citizens think that governments (or Google) are watching them at any given time, say, when they are searching the internet, then citizens are not intellectually free because they fear experimenting with new or deviant ideas. In this case, surveillance can chill one's exploration and creativity. Further, as stated, surveillance may disproportionately be directed at particular communities, which can impact the variety of creative contributions and viewpoints that make the world a better place. This disproportionate treatment also affects one's attitude toward privacy and surveillance in general (Gandy, 1993, p. 150).

In one more example of subtle effects, surveillance could also be tied to the feelings that people have for each other (and even societal and democratic attitudes), especially when it comes to social media. Surveillance is built into social media, whether it involves peers watching each other or platforms watching their users. Regarding peers, scholars have found that social media users curtail their posts based on how they want to manage their disclosure (Masur & Scharkow, 2016), and users think in terms of imagined audiences (Litt, 2012; Jung & Rader, 2016), who tend to be their social connections, leading them to limit what they post (Marwick & boyd, 2014). Further, in the case of the platform surveillance, platforms recommend content that can cause users to be less receptive to new ideas, and users presented with repetitive content biased toward their preexisting viewpoints may not realize how polarized their own views are (Dylko, 2017). Repetitious exposure to like-minded content can make people believe that everyone thinks the same way and affect their relationships towards others, predisposing them to be less tolerant.

Both effect and affect are particularly a concern for technical communicators because, as communicators, we are always asking about audience and effective arguments, and we are thus also asking one's attitude in relation to an argument. Further, the field is also concerned with how technology and information make someone feel. The whole area of human-computer interaction, usability, and user experience is concerned with affect. If surveillance and privacy issues have the potential for impacting one's experience, then surveillance (and privacy) issues should be a consideration with user experience. Further, for those interested in social justice, the felt experiences of surveillance are also connected to one's mental health, and how targets and even agents feel in situations of surveillance is a consideration worth exploring. Affect can thus be tied to the idea of chilling effects discussed previously, in that feelings can motivate behavior changes that could themselves be seen as harms. Even those like Snowden who conducted surveillance felt emotional anxiety carrying it out (Snowden, 2019a). It is worth mentioning here too, that, again, surveillance can be a positive thing, and some, such as those experiencing surveillance through social media, may feel empowered by the attention they receive; some technical communicators, such as those who write social media content, may also be happy to have more visibility or optimized searches online because that means they are doing their job well.

Resistance

The seventh thematic concern of surveillance that pairs nicely with TC scholarship is resistance. Surveillance is particularly difficult to escape from because it is often put in place through institutional and power structures. Governments, corporations, and those with authority—these are often the origins of systems of watching and recording, and to reject and escape them is no easy task. However, resistance can happen. One can tactically evade institutional practices through tactical communication, or one can take on the institution. Resistance can be small-scale, as when hiding one's face from a camera; or it can take place on a larger scale, when reacting not just to what has already been put in place, but to protests to the systemic foundations of systems of visibility, recording, and information storage in general. Resistance can be personal, or it can be social, for example, when rethinking a border's existence in the first place and not just thinking about a border's effect on you or others. This more longitudinal analysis of the various elements of surveillance is particularly complimentary to TC scholars and markedly addresses Walton et al.'s (2019) questions: "How can we recognize, reveal, reject, and replace oppression? What is the ideal we seek in its place? What does justice look like, and how is it enacted?" (p. 10). Surveillance allows TC scholars entry points into empirical spaces to "see surveillantly" to recognize, reveal, and reject oppression and countersurveillance. Resistance to surveillance can be a tactical eluding of structural power, or it can be an act of social justice.

Further, resisting surveillance would likely be incremental rather than punctuated equilibrium, and work to replace oppressive surveillance and putting something more just in its place would also mean embracing the "wickedness" of surveillance. The idea of a wicked problem, or a complex social problem without an easily definable solution, is also well known to TC communicators (e.g., Wickman, 2014), so this makes problem-solving more conceivable. Surveillance can seem overwhelming to address, particularly due to the power relationships, but if surveillance is viewed as a wicked problem, an entry point into making it more just and equitable starts a conversation as to what element of the wicked problem can be addressed to maximize impact. Research into wicked problems points out that some solutions are easier when the right questions are asked. While it might seem impossible to resist or rework an unfair system of surveillance, TC scholarship and TC scholars can assess situations of surveillance and apply their expertise toward asking the right questions to find solvable problems and advocate for vulnerable groups.

A Word about Privacy

Finally, eighth, TC can also offer help with resistance by decoupling surveillance and privacy. The connection between surveillance and privacy here is especially important for this book because it explains why the book mostly addresses surveillance, rather than surveillance and privacy. Privacy is crucial to understand surveillance, but privacy and TC is a topic for a separate book. Because this book focuses on linking a surveillance worker to a technical communicator, I want to stick more to the concept of surveillance and the agency found in resisting surveillance from the standpoint of a surveillance agent, rather than the more traditional story of resistance from the target's point of view. Resistance to surveillance is often paired to privacy, and the onus of privacy is put onto the target of the surveillance (Fernandez & Huey, 2009). Those under subject to surveillance must thwart the watchers. The target advocates for privacy protections to be solidified in legislation; targets maintains their privacy by limiting what they post on social media; they hide their faces so that they cannot be identified. The source being polygraphed puts a tack in their shoe (Marx, 2003).

It is easier to see why privacy and surveillance are linked by looking at the complexity of the words *surveillance* and *privacy*. The two words have different functions, but they are intertwined. One can "conduct" surveillance, implying that surveillance is something that can be performed. Privacy, on the other hand, tends to sound more possessed. We "protect" privacy. It is harder to imagine "conducting" privacy. We can conduct privacy checks or engage in privacy protections, but privacy itself isn't typically an action so much as a product. Although it is conceivable to "protect" surveillance, when used in this way, there is a base idea of protecting the process of surveillance rather than a potentially embodied noun of having surveillance to protect. One can have privacy as an individual possession or state of being, but one doesn't typically have surveillance.

However, while getting more privacy is useful for curtailing and thwarting surveillance, it is also limited. For one thing, as Coll (2014) notes, it is not wholly useful to situate privacy in opposition to surveillance because, in fact, it is often its ally. Privacy allows surveillance to be conducted because privacy facilitates secrecy. One needs privacy to surveil someone else. For instance, a voyeur is by definition someone who watches an unsuspecting other from a position of privacy.[10] Further, too, if privacy is what suffers due to surveillance, to lessen the impact of surveillance, one would need more privacy. However, to reduce surveillance, we don't just need more privacy. *Sometimes we need less surveillance.*

And that is where TC and its surveillant workers come in. Reducing surveillance is not just a matter of the watched taking precautionary privacy measures or the watcher paying more attention to the privacy of data of the watched; it is also a matter of the watcher taking into consideration where surveillance itself can be reduced. Resistance can then come from recognizing one's own actions as part of the surveillance process and evaluating it from a variety of cultural, just, and ethical standpoints. Surveillance thus matters not only because it has consequences for privacy; it also matters because it forces us to evaluate our own practices and those of others, to see where positions of power and visibility are established and executed. Often privacy hygiene becomes the way we resist surveillance, but this frame reduces our agency and culpability. As noted, surveillance can be understood as an everyday activity, and thus, every day, we may be the agent, not the target.

Final Opening Words

Overall, without a foundation in what surveillance is, it is harder to see the disciplinary assumptions the field holds. If we stick to panoptic models such as those that were popular in earlier, foundational literature and those popular to the surveillant imagination, we might not see the more everyday actions of surveillance that Staples (2000) might call microtechniques of control[11] or the "meticulous ritual of power," where knowledge-gathering techniques are "faithfully repeated" and "often quickly accepted and routinely practiced with little question," and which are "intended to discipline people into acting in ways" that others have deemed appropriate (p. 3). Chapter 2 moves on to address how technical communicators may come in contact with surveillance practices and begins to dig in deeper to the connections between surveillance and TC.

Chapter Two

Surveillance Workers and Technical Communicators

Moving more deeply into the discussion, this chapter asks who engages in TC and who engages in surveillance work? Whether or not it is immediately recognizable, I argue there is a range to both and that both can overlap. This requires an understanding of the technical communicator, surveillance worker, and a reconciliation of what it means to be doing both surveillance and TC. This chapter looks at technical communicators who do surveillance work, surveillance workers who engage in technical communication, and those who do not consider themselves primarily either surveillance workers or technical communicators but engage in either or both practices, whether formally or informally.

As explained in chapter 1, this book takes Lyon's (2001) position that surveillance is an everyday activity that involves "any collection and processing of personal data, whether identifiable or not, for the purposes of influencing or managing those whose data has been garnered" (Lyon, 2001, p. 2). Surveillance work, then, involves collecting, sorting, making sense of, and storing that information in a variety of everyday ways. However, despite automated and technology-assisted surveillance gathering processes, as this chapter will show, actions taken based on surveillance data also comes down to human intervention and the rhetorically constructed interpretation of what that data means, with surveillance workers, often serving as technical communicators (and vice versa), building and sharing these surveillance narratives. Edward Snowden becomes a useful touch point to illustrate the critical intersection of surveillance work and technical communication. Understanding these connections starts with a more in-depth look at the jobs held by Edward Snowden, the role of the technical communicator, and the role of the surveillance worker.

Snowden's Résumé

Snowden was part of the intelligence community in the US. At least seventeen agencies make up this community (Agrawal, 2017), and these range from the more well-known agencies of the Federal Bureau of Investigation (FBI), Central Intelligence Agency (CIA), and National Security Agency (NSA), to the less pop-culturally dramatized agencies like the National Geospatial-Intelligence Agency (NGA), Defense Intelligence Agency (DIA), or the National Reconnaissance Office (NRO). The CIA and NSA offer two illustrations of surveillance work that agencies engage in. Among many other job areas, the type of work contributing to the overall mission at the CIA spans from analysis jobs, clandestine operations, STEM- and computer-related activities, support jobs, and foreign language occupations (Central Intelligence Agency, 2018). Job types at the NSA range from intelligence analysis and collection, foreign language analysis, computer science, cyber, engineering and physical sciences, mathematical science, and business and accounting (National Security Agency, "Careers," n.d.). A purpose of these agencies is to conduct activities pursuant to Executive Order 12333, which calls for "the necessary information on which to base decisions concerning the conduct and development of foreign, defense and economic policy, and the protection of United States national interests from foreign security threats" (1981). The community works to "enhance human and technical collection techniques" to carry out their goals. In essence then, for their economic livelihoods, those in the intelligence community operate, obtain, interpret, and use surveillance information to direct actions and policies of various agencies.

While Snowden's résumé isn't widely published online, according to intelligence officials, Edward Snowden was classified as a systems administrator (Drew & Shane, 2013) for the NSA. The U.S. Department of Labor (DOL), Bureau of Labor Statistics outlines that network and computer system administrators work with the "day-to-day operation" of their organization's networks (U.S. Department of Labor, 2018c). Shane and Sanger (2013) add that it "is a bland name for the specialists who keep the computers humming," and using the workers' "store[s] of technical manuals," these are "the technology workers with the most intimate knowledge of what is moving through their employers' computer network" (Drew & Sengupta, 2013).

As Tina Darmohray at Short Topics in System Administration (SAGE), a professional organization for systems administrators, reports, the best employees "can wire and repair cables, install new software, fix bugs, train users, offer tips for increased productivity across areas from editing to databases, evaluate new hardware and software, automate a myriad of mundane tasks, and expedite work flow at their site" (2010, p. 1). They also "gather data to identify company needs" such as security measures they could further use "to determine and evaluate system and network requirements" (Association for Career & Technical Education, 2005, p. 40).

As a contractor with both Dell and Booz Allen, Snowden appeared to do a wide range of these duties in this role. As per sources who had viewed Snowden's résumé, he reportedly "was responsible for the security of 'Windows infrastructure' in the Pacific" and "he led a project to modernize the backup computer infrastructure" (Drew & Shane, 2013). Snowden (2019a) professes to have been involved with connecting the NSA's systems architecture to the CIA's and "being the only one in a room with a sense not just of how one system functioned internally, but of how it functioned together with multiple systems—or didn't" (p. 166).

Working from Hawaii (Harding, 2014), in the day-to-day labor, Snowden would have worked with extensive freedom to move about systems without auditing of his actions, because, as Esposito and Cole (2013) report, Snowden could be considered *the* audit. As computer security company president Eric Chu commented, "The system administrator has godlike access to systems they manage" (Drew & Sengupta, 2013), and Snowden was the one tasked with making sure files weren't corrupted and systems were running as planned. This afforded him "the right to copy, to take information from one computer and move it to another," and he could basically look at any file he wanted "without leaving any signature" (Esposito & Cole, 2013). Further, it was even important for him to move files, because "Snowden actually had the responsibility to go to the NSA intranet site and move especially sensitive documents to a more secure location" (Gjelten, 2013).

Reportedly on his résumé, more recently Snowden had gained ethical hacker certification "by studying materials that have helped tens of thousands of government and corporate security workers around the world learn how hackers gain access to systems and cover their tracks" (Drew & Shane, 2013), and had moved into different work he described as being an "infrastructure analyst" (Shane & Sanger, 2013) (although the NSA still publicly listed him as a systems administrator). Snowden himself reported working as a systems administrator (Snowden, 2019a, p. 135). This constituted a shift from defensive work to offensive work "in which the N.S.A. examines other nations' computer systems to steal information or to prepare attacks," and is a common career pattern (defensive to offensive) often encouraged by the NSA (Drew & Shane, 2013). Doing these duties, he would have been in more of a position "like a burglar casing an apartment building," who "looks for new ways to break into Internet and telephone traffic around the world" (Shane & Sanger, 2013). Supposedly, too, the résumé noted that he went from "supervising computer system upgrades for the spy agency in Tokyo to working as a 'cyberstrategist' and an 'expert in cyber counterintelligence' at several locations in the United States" (Drew & Shane, 2013). One "perk" of these duties was that Snowden was able "to gain access to lists of computers that the agency had hacked around the world."[1] This is characteristic of infrastructure analysts, because, as Shane and Sanger (2013) continue: "Infrastructure analysts like Mr. Snowden, in other words, are not just looking for electronic back doors into Chinese computers or Iranian mobile networks to steal secrets. They have a new double purpose: building a target list in case American leaders in a future conflict want to wipe out the computers' hard drives or shut down the phone system."

Snowden as a Surveillance Worker— and Who Are Surveillance Workers?

Now drawing from surveillance studies, Snowden, in his duties for the intelligence community, fits the role of the surveillance worker. According to Gavin J. D. Smith (2012), surveillance workers (or those I will later classify as capital S, capital W Surveillance Workers) are "data-deciphering 'optometrists'" who are "responsible for the daily operation of surveillance" and who "contribute their interpretive energies in exchange for the accumulation of tacit knowledge, emotional stimulation and a salary" (p. 107). These individuals also transform the information they obtain into actionable intelligence to assist whatever organization they work with. Smith continues that they "attribute both the products of their labor and the wider manufacturing process with classificatory codes and meaning" and engage in "the art of monitoring, interpreting and making sense of social reality" (p. 107), to try to understand "unknown futures" (p. 108). Surveillance workers thus operate in both the known and the unknown, rhetorically constructing arguments as to what data they gather might mean and influencing how those with the power to make decisions should respond to their constructions, using both their own interpretive skills, knowledge of social norms, and operating within bureaucratic confines of what is allowed by internal policy, law, and budgetary constraints. Their actions both seek to break down "cultural conventions, rules, norms and customs," as well as attempt "to meet a variety of bureaucratic ends and ideals" (p. 107). Their work is rhetorical and blends information from both human input as well as information from the "nonhuman labor force" from sources such as technology for "'reading,' categorizing and deciphering this data" to eventually create "intuitionally-relevant knowledge" (p. 108). As no one can predict the future, the best a worker can do is try to interpret the past, mesh it with the present, and try to guess what is to come. In Snowden's work with surveillance information, as previously reviewed, Snowden would have engaged in interpreting data maintained in files on systems he worked with in exchange for a salary for that specific work.

Along the range of surveillance workers lies a degree of surveillance responsibilities. As Smith (2012) points out, the "type and intensity" of surveillance varies depending on "the precise aim and function of the observation," as well as "whether such practices have significant implications for the subject's mobility or well being" (p. 109). An FBI agent watching a suspected embezzler whom the agent could ultimately assist in sending to prison is a different type and with a different intensity from a social media influencer following another influencer's European vacation social media posts to assess business strategies (a range that will be explored as surveillance in chapter 3). Both exhibit characteristics of surveillance in that someone is watching someone else to gain more insight, but the FBI agent has decidedly more intensity and implications for eventual mobility. Thus, there is a distinction between a "Surveillance Worker" and a "surveillance worker." Figure 2.1 illustrates a possible spectrum with different positions in an auditing field, ranging from personal accounting (which could have close surveillance connections but is being used here to represent the accounting of an everyday individual not participating in overly surveillant practices) to FBI forensic accountants actively engaged in monitoring the accounts of a suspect. The range moves from a smaller scale to larger corporate practices of oversight, to the government targeting a suspect. While any job could move along the continuum, depending on job duties, this figure represents a moment of time.

On a range, capital S, capital W Surveillance Workers engage in practices of surveillance for a living, and lowercase s, lowercase w surveillance workers carry out surveillance but aren't occupied full time carrying out surveillance. Those employed as the "Surveillance Workers" fit the more traditional roles of surveillance or those of the surveillant imaginary (or how media stereotypically might imagine surveillance to be through cultural presentations like TV shows or movies), who get their paycheck from working with surveillance-obtained information and are in positions where the surveillance work being carried out has the power to significantly affect the lives of those being watched. Traditional jobs in the surveillance imaginary like those in the intelligence community, government workers of Big Brother in *1984*, and the prison guards in the panopticon fit this description—Surveillance Workers, authorized, capable, and armed with surveillance techniques, gathering information on the watched in return for a salary to control populations or influence with power.

Those employed as *s*urveillance *w*orkers on the other end of the spectrum, would carry out surveillance practices and get paid for their work, but their actions would have less significant implications for mobility and well-being of the watched, with a reduced intensity from work higher up in the range. The watching also might be because of "everyday surveillance," as discussed, rather than more formal definitions of surveillance along power and institutional lines. As Smith (2012) continues, basically, "any occupation or trade requiring an employee to collect, assess, and process information—and action particular responses based on that data—is in effect asking that individual to conduct surveillance work" (p. 108), so,

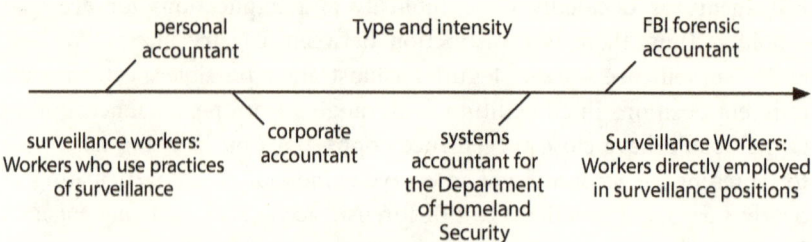

Figure 2.1. A visual of the range of the surveillance worker. Author provided.

this places several lower-surveillance-intensive positions on the range. This could be an insurance agent, bank teller, accountant, or technical worker tasked with gathering information about individuals—anyone using surveillance practices through the course of their occupations, despite not fitting into the surveillance imaginary more represented by the Surveillance Worker. These types of surveillance workers are increasingly more prevalent, as concepts of what surveillance is are expanded. No longer relegated to narrower and more focused monitoring left to police or security officials in reference to crimes, such as traditional definitions like that of McQuade and Danielson (2005), everyday surveillance becomes something more widespread and pedestrian, reducing or at least altering where we can identify and claim spaces of surveillance. If surveillance exists in the everyday, then surveillance workers are also likely to be more "everyday" surveillance workers, rather than Surveillance Workers existing at the extreme and professional end of the spectrum.

Who Are Technical Communicators?

Technical communicators similarly exhibit a range of job duties, which can ultimately overlap with responsibilities of S/surveillance W/workers. While it is admittedly complex to answer the question of who a technical communicator is, and a definitive answer is neither desirable nor my goal here, for this book, I argue that technical communicators can also exist on a range, particularly when looking at the range exhibited in the contrast between a definition from the Society for Technical Communication (STC) (2018) and the work of Henning and Bemer (2016). Both works are in no way comprehensive of the whole field but do offer a comparative snapshot of two varying positions. A more comprehensive look at who a technical communicator can be found in TC literature such as Kimball's (2017) description of TC as an activity, mentioned in chapter 1, or other sources—and I encourage a review of that literature for further analysis.

Back to the argument, the STC gives one portrait of a technical communicator. According to the STC, TC is a broad area that encompasses many jobs, but it is held together with communication that has at least one of the three characteristics: (1) the communication relays "technical or specialized topics"; (2) the communication communicates "by using technology"; and/or (3) the communication "provides instructions about how to do something" (Society for Technical Communication, 2018).

Technical communicators, then, could be seen as those doing these activities. Some sample occupations the Society lists are technical writers and editors, indexers, information architects, globalization and localization specialists, usability and human factors professionals, visual designers, and web designers and developers. Regardless of their differences, the STC affirms that for those working in the field of technical communication, "What all technical communicators have in common is a user-centered approach to providing the right information, in the right way, at the right time to make someone's life easier and more productive." They do this by advancing the goals of their employers by making "information more useable and accessible to those who need that information" (Society for Technical Communication, 2018).

A second argument is from Henning and Bemer (2016) who look at what being a technical communicator means, not just according to the work they do but according to the training received and skills they can offer. They argue that a strong definition of a technical communicator provides for what (Carliner, 2012) might call "brand identity" and provides communicators the ability to show others what it means to be employed in this type of position by helping define the work that they do. They assert that a stronger identity is necessary because without a definition, "members of the field lack a unifying way to present themselves to employers and collectively argue for their value and the value of their work," and "technical communicators are left in a position where their work may be encroached upon by others who lack their specific knowledge and expertise," which lowers "the value of technical communication deliverables and processes" and ultimately hurts "the legitimacy of technical communication as a field" (p. 313).

Building on a U.S. Department of Labor, Bureau of Labor Statistics 2012 definition of a technical communicator (but accounting for more skill diversity and flexibility), Henning and Bemer (2016) support the revised definition of a technical communicator as follows: "Technical writers, also called technical communicators, produce documents in a variety of media to communicate complex and technical information. They employ theories and conventions of communication to develop, gather, and disseminate technical usable information among specific audiences such as customers, designers, and manufacturers" (p. 328). In forming this definition, the technical communicator acquires a definable, established position with clearly defined skills offered to potential employers.

In asserting this definition, too, the teaching of technical communicators gets strengthened, because a classification helps empower academia to produce workers prepared for these types of positions. If a job description is identified, then scholarship can reflect related practical and conceptual skills required to meet expectations of these professions. As Henning and Bemer (2016) continue to explain, "Scholars seem to at least agree that academics and practitioners need to work together if they hope to professionalize the field with a common body of knowledge and thereby gain power and legitimacy" (p. 321). Henning and Bemer (2016) further find that technical communicators need the practical skills of editing, writing, and technology, and as Henschel and Meloncon (2014) note, they need the conceptual skills of "rhetorical proficiency, abstraction, experimentation, social proficiency, and critical system thinking" (p. 10). Schools, then, can be reassured that the curriculum being presented prepares students for the roles they will be entering and thus, a definition also helps unite the academic training before a candidate enters the workplace.

Looking at the two discussions, Henning and Bemer (2016) make an argument that supports a tighter, more defined description of a technical communicator than the STC (2018) who noted that TC is broad and includes any one of the three noted three characteristics: communicating technical and specialized topics, communicating using technology, and providing instructions (para. 1). On the other hand, Henning and Bemer's argument is that a more specific definition involving training and skills helps create power in the position to "bring legitimacy to the field" (p. 312). While the STC's (2018) definition also provided specific descriptions of the work being done by technical communicators and descriptions of the work being done, it does allow for broader interpretation of who a technical communicator may be based on engaging in characteristics of technical communication. As described in their definition, if just one characteristic is present, an individual can be engaged in technical communication. For instance, if one communicates "by using technology," then they can be acting as a technical communicator.

The nuances between these definitions, though, open space to talk about surveillance, surveillance workers, and their intersection with TC. Adhering more to the STC's definition that offers more leeway in description, but still thinking about Henning and Bremer (2016) and their narrowed definition, the juxtaposition clears a space to see that some individuals may conduct the job duties of a technical communicator while not engaging in occupations typically indexed to be technical communicators. I would argue then, that, as with the S/surveillance W/worker, there is a range in what can be considered a technical communicator. Those considered with the capital *T* and *C* serve in more formal positions of Technical Communicators, having received specialized training in technical communication or and having jobs that explicitly revolve around formally communicating technical information;[2] those technical communicators labeled with a lowercase *t* and *c* work technically, communicating by exhibiting at least one of the three characteristics discussed by the STC, but they work in other positions that do not predominantly focus on being a "Technical Communicator." Figure 2.2 illustrates this concept.

For instance, system administrators may communicate to bosses or stakeholders about the technical aspects of the work they are doing and could thus be engaged in communications that relay "technical or specialized topics" (Society for Technical Communication, 2018), but they aren't practicing in a role that would immediately be considered as "Technical Communicator" and may not have specialized training in technical communication or need/want this specific training because most of their time is spent organizing and developing computer systems. Technical writers, on the other hand, are more easily identifiable as *Technical Communicators*, because they are usually defined as being in that role and likely have some type of training that prepares them for the associated job tasks. A job's correlation to the continuum ranges according to the intensity of duties and requirements involved, in the context of TC practices.

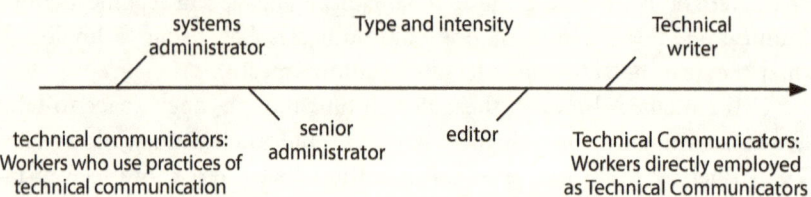

Figure 2.2. A visual of the range of the technical communicator. Author provided.

Ranges and the Matrix

What emerges from these ranges of distinction for both technical communicators and surveillance workers is a larger, more complex matrix of at least four areas. Figure 2.3 presents a matrix of positions and begins to sketch the spaces where technical communication work overlaps with surveillance work. For those studying technical communication, this is important to consider because, if one views jobs on a range, TC doesn't always take place in spaces confined to the Technical Communicator.[3] Without considering these ranges in the two occupational areas, it is harder to conceptualize the overlap between the two concentrations. A technical writer would have to work directly in some type of surveillance position to see the connection (e.g., a technical writer at an intelligence agency). But really, issues affecting technical communications also occur in communications carried out by those engaged in aspects of TC, rather than just by those directly employed in TC positions.

To understand this matrix, it is helpful to examine each quadrant, starting with the area where Edward Snowden fits, looking at both Snowden and the job in which he served in general. Snowden provides a strong example of the Surveillance Worker/technical communicator area in the lower-right area of the matrix. As discussed at the beginning of this chapter, while doing his work with the NSA, Snowden was employed in several technical jobs not expressly considered positions

Technical Communicator	E.g., User experience researcher (TC/sw)	E.g., Technical Writer II with the Marine Crop (TC/SW)
technical communicator	E.g., Accountant (tc/sw)	E.g., Systems administrator with N.S.A. (tc/SW)
	surveillance worker	Surveillance Worker

Figure 2.3. A matrix of technical communicators and surveillance workers. Author provided.

of a "Technical Communicator." One of those positions was a systems administrator. While working with surveillance data, he would also have had to engage in communication involving the STC's three characteristics of technical communication. For instance, he would have practiced the STC's characteristics of technical communication when (1) relaying "technical or specialized topics" to others, while "supervising computer system upgrades for the spy agency in Tokyo" (Drew & Shane, 2013); (2) communicating "by using technology" during his time working remotely in Hawaii (Harding, 2014); and (3) providing "instructions about how to do something," while he was engaged in a leadership role when "he led a project to modernize the backup computer infrastructure" (Shane & Sanger, 2013). Thus, although working categorically as a surveillance worker, he also would have exhibited practicing technical communication, though not employed as a Technical Communicator.

These duties are also common when looking at the position overall and not just at Snowden, as any systems administrator would have to carry out technical communication work. These duties, in comparison to the work of a technical communicator, illustrate how a systems administrator can be doing technical communication work. The definition of a technical communication through which to compare the systems administrator position with is taken from the Society for Technical Communication's (2018) web page, "Defining Technical Communication," but it is a disclaimer that, as this book has and will illustrate, what it means to be a technical communicator or do technical communication work is not limited to a three-point definition.

First, they would have to communicate "technical or specialized topics" (Society for Technical Communication, 2018). The U.S. Department of Labor (2018b) even lists "communication skills" as one of the important qualities for this job, because "[a]dministrators must describe problems and their solutions to non-IT workers." In Darmohray's (2010) description of the role, the first *required* skill listed for any systems administrator, from novice to chief information officer, is "[s]trong interpersonal and communication skills." Further, depending on the level of work individuals are doing, in the course of their technical work on their systems, they need to have the ability "to explain simple procedures in writing or verbally" (p. 3); "to write basic documentation" (p. 4); "to write purchase justifications" (p. 5); to "make presentations to an internal audience" (p. 6); to "interact positively with upper management" (p. 6); "to write proposals or papers, act as a vendor liaison, make presentations to customers or client audiences or professional peers" (p. 6); to have "[s]trong writing, meeting, and organizational skills" (p. 9); to have "[w]riting, presentation, negotiation, facilitation, meeting, and organizational skills" (p. 11); and they need to have a "[b]ackground in technical publications, documentation, or desktop publishing" (p. 15). All these skills would essentially be communicating technical topics, but as a systems administrator typically requires a degree in computer systems (Association for Career & Technical Education, 2005; Darmohray, 2010; U.S. Department of Labor, 2018b) and not TC, specialized technical communication may be (or may not be) absent from the worker's background. Knowledge of these skills may come from apprenticeship work, improvisation, or related coursework, but the skills aren't as likely, although it is not impossible, to be obtained through a technical communication degree, because the education requirements include computer or information science degrees.

Second, workers would also be "Communicating by using technology, such as web pages, help files, or social media sites" (Society for Technical Communication, 2018). Communicating through technology is arguably broad as defined there, but also a requirement for a systems administrator and one that would show up in two different ways. From a stance that sees coding as a type of writing or literacy (Beck, 2015; Cummings, 2006; Easter, 2018; Gallagher, 2017; Haefner, 1999; Hope, 2002; Lindgren, 2021; Messina, 2021; Vee, 2017; SIGWROC, n.d.), communicating using technology is a required skill for a systems administrator. As Darmohray (2010) discusses, desirable qualities for this position are the "[a]bility to write scripts in some administrative language (e.g., Perl, Python, VBScript, Windows PowerShell)" (p. 5), the "[a]bility to port C programs from one platform to another and to write small C or C# programs" (p. 7), or having a strong understanding of programming languages (p. 11). System administrators would be engaging in communication practices of writing, coding, and developing programs to complete their jobs. Also, in a more traditional sense, as described above, administrators would also need to communicate with others to do their jobs, tasks that would inevitably involve communicating through technology. For instance, those in this position would have a desirable "[b]ackground in technical publications, documentation, or desktop publishing" (p. 15).

Third, workers would need to engage in communication that "provides instructions about how to do something" (Society for Technical Communication, 2018). As described in the job requirements, another part of the job involves supervising and teaching others at any range of the job level. For instance, the U.S. Department of Labor (2018e) lists that those in these positions "[t]rain users in the proper use of hardware and software." Darmohray (2010) goes more in-depth and breaks this teaching down into various levels. For instance, in the first tier of job ranks, junior administrators would need to have the "ability to train users in applications and operating system fundamentals" (p. 4); and intermediate/advanced employees would be involved in "training users in complex topics" (p. 5) as well as managing "novice system administrators or operators" (p. 6). Senior positions would serve as a "technical lead and/or supervises system administrators, system programmers, or others of equivalent seniority" (p. 7). In what Darmohray categorizes as management positions, technical leads would help "the system administration manager in setting staff goals and training, defining technology priorities, and developing long-term strategies to manage and scale system administration" as well as managing "one or more staff" to give them "technical guidance and mentoring" (p. 8); and the system administration manager would have the "[w]illingness to mentor, train, and share knowledge with team members" (p. 9), as well as supervise others for "technical guidance and mentoring," and would need to give "career guidance and performance feedback to team members (p. 10). The IT director "[s]upervises one or more direct managers and provides them with tactical guidance and mentoring," also giving "career guidance and performance feedback to direct reports" (p. 10); and the chief information officer would be able to manage "one or more direct managers and provides them with strategic guidance and vision," as well as give "career guidance and performance feedback to direct reports" (p. 12). All these tasks provide information to others about how to do something, even if just improving one's performance.

In all positions, too, administrators would also share the technical communicator's goals of creating a "user-centered approach to providing the right information, in the right way, at the right time to make someone's life easier and more productive" (Society for Technical Communication, 2018). The job duties revolve around streamlining processes and making organizations more productive by optimizing systems and training users to use the systems correctly (U.S. Department of Labor, 2018e). Thus, this shows that those holding similar positions, but not traditional Technical Communicators, can engage in practices of technical communication without being *Technical Communicators*.

While it may not be a stretch to call an engineer a technical communicator, Snowden and the systems administrator, however, only represent one quadrant, and looking at the other quadrants are just as important. Moving around the matrix, the next area in the lower-left quadrant is the technical communicator/surveillance worker (tc/sw). In the surveillant (and TC) imagination, this area probably represents the least stereotypical overlap between surveillance and TC. Jobs in this quadrant don't immediately come to mind as either TC or necessarily surveillance jobs. But, if classified on a range, the position exhibits qualities of both. For instance, going back to the accountant example, while the job description for an accountant doing personal taxes might not immediately read as Technical Communicator or Surveillance Worker, looking at the job description, traces of both pop up.

At first glance, an accountant wouldn't be categorized as a Technical Communicator in the Henning and Bemer (2016) sense, as this person would be a trained accountant focusing on revenues, costs, and taxes, not functioning in an organization to produce the same sorts of documents as a Technical Communicator with rhetorical training. As the U.S. Department of Labor describes, an accountant (or auditor) would "prepare and examine financial records" to "ensure that financial records are accurate and that taxes are paid properly and on time" to "help ensure that organizations run efficiently" (U.S. Department of Labor, 2018a). This is not really the domain of technical communication. However, the accountant does engage in practices of technical communication. According to the U.S. Department of Labor (2018a), "In addition to examining and preparing financial documentation, accountants and auditors must explain their findings. This includes preparing written reports and meeting face-to-face with organization managers and individual clients." Thus, they must create documents synthesizing specialized information (most likely using some type of technological assistance) to help others understand what was carried out.

On the surveillance side of things, if people take their financial information to an accountant to determine their tax debts, they don't necessarily look at this individual as a surveillance worker, most likely because they requested the accountant's work and submitting the paperwork doesn't significantly impact their mobility or well-being. But the connection between surveillance and accounting has been explored by Macintosh (2006), and this individual would be engaging in Lyon's (2001) definition of surveillance in the everyday sense by engaging in "any collection and processing of personal data, whether identifiable or not, for the purposes of influencing or managing those whose data has been garnered" (Lyon,

2001, p. 2). By reviewing the financial information on someone or someone's business, even if the information is self-provided, the accountant would be working to help influence what should be submitted for taxes or managing their finances. So, while an accountant may neither fit the description of a Technical Communicator nor a Surveillance Worker, that person can be undertaking less formal work of both, and thus shows up in the matrix.

Around the matrix, there are also individuals doing more traditional Technical Communicator work but not necessarily the traditionally conceptualized Surveillance Worker (that is, a surveillance worker before considering the range and scope of a Surveillance Worker), who would fall into the area of Technical Communicator/surveillance worker (TC/sw). An example of this person would be someone engaged in user experience work. For instance, a firm providing software and data analytics to real estate companies ("Company," 2018), posted the job description "User Experience Researcher," which mainly focuses on job duties of a Technical Communicator ("Careers," 2018). Technical communication-related activities in the post involved the abilities to "[a]nalyze, synthesize and communicate research insights and findings to the stakeholders and be the influencer for change," as well as do this through technologies like the "UX design tools" of "Sketch, Photoshop, etc."

While this position seems like it could be directly categorized as a Technical Communicator, the surveillance portion of the job is less clear. Surveillance-wise though, the researcher would be analyzing user experience to form their conclusions, a type of oversight used to collect information on others to manage or influence others. These activities include the need to "[p]lan, design and execute" both qualitative and quantitative studies ranging from field observations, to interviews, in-person studies, surveys, to a/b tests in order "to turn insights from research into solutions to business requirements." Thus, they would be gathering surveillance information to create actionable organizational goals. In this intersection then, positions like this are primarily focused on Technical Communication, but they do employ activities recognized as forms of surveillance. Because of this, I argue their category is TC/sw.

The fourth and final quadrant of the matrix sits the Surveillance Worker/Technical Communicator section. A job in this area would be firmly rooted in both TC and surveillance, and an example illustrating this is a Technical Writer II job advertisement with a contractor for the

Marine Corps Intelligence School in Quantico, VA. According to the ad, regarding TC skills, the desired individual would be involved in creating and editing technical and policy documents for the capabilities, procedures, methods and configuration associated with auditing and insider threat programs. as well as work on documents "to convey technical material, business processes, and policies in a concise and effective manner" (Indeed, 2018). For surveillance skills, the individual would need "Five (5) years of experience with Expeditionary Intelligence Programs" and procedures and guidelines for C4ISR [Command, Control, Communications, Computers, Intelligence, Surveillance and Reconnaissance] systems or equipment." In this instance, there is a strong overlap between both TC and surveillance work. Evaluating the job ad, the individual would be creating technical documents involving surveillance and intelligence systems for the purpose of gathering or processing information on individuals to manage them.

The four quadrants illustrate that both TC and surveillance work can take place on a range, and while some jobs look strongly rooted in either Surveillance Workers or Technical Communicators, other positions lie more along a continuum. As this book continues, it will thus assume that both activities take place on a range.

The Overlap's Social Implications

This conversation is especially crucial to have where the positions begin to overlap. As shown by the situation around Snowden and subsequent privacy implications discussed in chapter 1, Snowden shared that the US was conducting surveillance on not just suspected law breakers, but on everyday US citizens as well and world leaders. Snowden, although for the most part a Surveillance Worker, played his part in the surveillance scenario[4] by carrying out duties shared by technical communicators by relaying technical topics, communicating using technology, and providing instructions on how to do things during his day-to-day work. Although his background was not in TC, as shown by his résumé and the job duties of a systems administrator, he likely would have engaged in work associated with TC.

So, although TC programs may not have seen a Snowden or other similar majors in their courses, because Snowden was on the less intensive end of the TC spectrum where his job might not require a background in TC (for instance, Harner and Rich [2005] found, only 2.5% of technical communication programs are located in engineering schools, although most assuredly that doesn't mean an engineer hasn't taken TC courses, as this area is often a consideration of TC). On the flip side, TC programs may see plenty of Technical Communicators who fit the low end of the Surveillance Worker spectrum and engage in of practices of surveillance. As Smith (2012) called for, as scholars we need to consider how surveillance labor is "enacted, performed, managed, regulated, experienced and understood" (p. 109), and this includes many sectors of employment, including fields of TC where workers are often at the frontlines of policy and technology. Not every surveillance worker is out working for the FBI in an intelligence community; some Technical Communicators as well as technical communicators are out conducting surveillance practices in the everyday surveillance spaces, for example, by carrying out activities like surveillance writing (as chapter 6, in particular, discusses in more detail), and it is important to identify these spaces, especially as information communication technologies (ICTs) with surveillance capabilities become more and more prevalent. This connection also allows for a more interdisciplinary conversations that illustrates how almost anyone using technology can be a technical communicator, in some sense of the word, and also encourages scholars to look at technology users as technical communicators, which puts the scholarship of TC directly in conversation with disciplines such as media, communication, information, sociology, and computer science.

In a more practical sense, scholars of TC are teaching students to go out into the world and practice it, but in an era increasingly dominated by surveillance and everyday surveillance practices such as government surveillance, big data analytics, monitoring of user experience, and social media analysis, it is increasingly important for students to understand both the convergent practices of surveillance and the repercussions of this. Students would benefit from understanding surveillance practices and examining their ethical stances and limitations before venturing into fields requiring that type of work to be accomplished, and understand spaces of social justice through which they can resist themselves or help others when surveillance seems unethical. Chapter 4 will discuss further why surveillance matters, and chapter 6 will further explain discussing surveillance with students.

Possible Resistance

Broadening the understanding of a technical communicator is a strategic endeavor, one that takes into account both the work that a technical communicator does as well as the scholarship that comes out of the field. Similarly, increasing the idea of what it means to be a Surveillance Worker might also be problematic in that it becomes too easy to see surveillance everywhere. But while it is strategic to broaden the two groups of people, it can also be viewed as necessary. For TC, Henning and Bemer's (2016) definitional argument still leaves room for surveillance workers doing TC work, if one also looks at their inclusion of flexibility. They comment, "This definition's existence does not mean it cannot be adapted to fit a particular career or situation (after all, not all medical doctors practice the same type of medicine) but it does mean that members of the field have something to point to when others doubt their worth" (p. 335). That the technical communicator is not only defined by duties and training isn't a novel idea, however, and Henning and Bemer (2016) were in part reacting to these more inclusive definitions in the first place. TC scholarship is expansive in what it means to do TC.

What Comes Next?

Overall, taking a closer look at both surveillance work and work in TC, the workers that carry out characteristics of these areas can either be directly employed in positions specific to the field, or they can work in other areas and carry out corresponding duties. It is important to enlarge the frame of reference in both areas, because there are quadrants of people doing surveillance work that involves technical communication who wouldn't traditionally be considered Technical Communicators or Surveillance Workers, but who are engaging in both areas. These areas represent rich areas of exploration for TC scholars to delve deeper into processes of surveillance and the ramifications of the processes and policies. The following three chapters start to interrogate further the connections between TC and surveillance, especially with information, evaluation, and resistance, and discuss more themes present in Snowden's example to broaden the conversation.

A value to using this lens is that if we think of technical communication in a broader scope, we can see that at any given point *anyone* could be a technical communicator engaging in communication involving technology. But just as not just everyone can or should teach writing, and it requires training and expertise to be a writing teacher, Technical Communication is a specific profession. The benefit of thinking of TC in upper- and lowercase terms then is that we can separate a more casual, albeit ubiquitous idea of technical communication from those fields requiring additional expertise.

Chapter Three

Information, Technical Communication, and Surveillance

Building on the last chapter, which discussed the range of technical communicators and surveillance workers, this chapter will look at the underlying element connecting both elements: information. To communicate in a technical way, and to engage in practices of surveillance, one must work with some type of information.[1] For this chapter, there are two questions that need to be addressed. The first is how does a technical communicator work with information? Second, how would one begin connecting this information to surveillance? To answer with the broadest reach requires a conversation regarding what exactly technical communicators do, and also requires a better idea of what surveillance information can be. The chapter overall argues that technical communicators work with a variety of surveillant information, and a useful way to "see surveillantly" (Finn, 2012) and identify how information is surveillant is to use one's surveillance glasses, so to speak, to see the elements of surveillance in any surveillance scenario.

Information as Key to Technical Communication and Surveillance

Due to the ubiquity of the word, *information* is often taken for granted as an understood concept rather than as a concept for inquiry. Ramage and Chapman (2011) note that there is an assumption that "information is something tangible, the equivalent of a physical object that can be stored and processed within an information system" (p. 4). Jaeger and Burnett (2010) comment that so often academic fields ignore a theoretical look at information, which can leave information "as a simple and unambiguous signal passing through the conduit of communication systems" (p. 7). Even when a given field doesn't take the word for granted and explore the term, *information* has contested meanings, and it is used by a variety of disciplines in various ways. These assumptions are so pervasive that Ramage and Chapman (2011) conclude, "There is no a priori reason to suppose that the word means the same thing when used in different contexts." Further, with competing concepts, "a single definition of the word might be impossible" (p. 4). For this chapter, *information* will be simplified as "*data* plus *meaning* (interpretation) in a particular *context* at a particular *time*" (Holwell, 2011, p. 72) (emphasis in original).

For the purpose of this discussion, this chapter will go back to the STC's definition of TC, which states that technical communicators work to "make information more useable and accessible to those who need that information" (Society for Technical Communication, 2018). This is useful for laying out, first, how technical communicators use information and, second, for identifying where surveillance intersects with TC. Hall and Wahlin (2016) also make a technical communicator's connection to information by adding, "Ultimately, the goal of technical communication is to transmit important information as effectively and efficiently as possible—information that allows you and the people around you to do your jobs well."

To make information more usable, accessible, effective, and efficient, a communicator employs principles of rhetoric. As already discussed in chapter 1, there are many dimensions to the word *rhetoric*, and I argued that surveillance itself was an argument/rhetoric. Returning to that discussion a bit more, it's useful to shift directions in rhetorical scholarship to look at how rhetoric plays a role in the structure of communications, particularly through the lens of TC scholarship.[2] In a general sense of the connection, Wysocki (2013) argues that rhetoric shows us "how any communication does what it does" (p. 441). Porter (2013) argues that rhetorical theory "is indispensable to technical communication" and provides the framework for everything, from documents to design, to the writing process, research, methods, interpersonal communication, and audiences (p. 141). Wysocki further states that rhetoric helps analyze text, think about strategies, and consider audiences, contexts, and purposes for the communication.

Classic tenets of rhetorical theory that communicators can pull from range from ancient rhetorical appeals of ethos, pathos, and logos; to the five canons of rhetoric, which are invention, arrangement, style, memory, and delivery; to three branches of rhetoric—deliberative, forensic, and epideictic rhetoric (Meyer, 1996). Communicators also consider the audience with ideas of context, situation, or cultural setting (Thatcher et al., 2011), and they consider the way images or other multimedia elements offer another level of audio, visual, and digital affordances (Kostelnick, 2019). Johnson-Sheehan (2002) argues that a communicator must first assess a situation using rhetorical strategies through a process called "interpretation," but then the communicator executes strategies based on that interpretation in a process called "expression," or the performance of a rhetoric.

Further, rhetoric is not just a way to teach or think about creating and arguments, rhetoric becomes a method for breaking arguments apart to see what type of persuasion is going on and for theorizing and speculating about what an audience is or how an argument might be perceived in its cultural context. For instance, rhetorical analysis can look at the ways particular phrases, language, or terms are used in particular contexts (e.g., Lawrence et al., 2019), or it can provide a way to look at how language and artifacts work together to build new insights and narratives (Cheek, 2020).

Essentially, one way a technical communicator works with information is by sorting it into the most relevant and appropriate chunks most suited for a particular audience by identifying relevant information, appealing to certain values, and adapting a message both visually and content-wise for a particular situation to maximize the effect of a certain message. One of the technical communicator's key skills is understanding the rhetorical importance of delivering technical or specialized information.

In surveillance studies, information is often approached in similar frameworks, albeit in different words. Data is often what is being surveilled, but information is about what is interpreted about it. Data can be someone's credit history, but information might be the interpretation of that data. Information is what has been done to this data to make it make sense.

Information in surveillance scenarios is thus also rhetorical, relating to the ways data is analyzed and transformed into actionable, decision-making chunks by being sorted into categories of risk or reward. In surveillance, *sorting* is used as the word to describe the process of selecting data to turn into information to deliver to the right audiences at the right times. Sorting is a key element of surveillance, and contemporary surveillance practices are driven by categorization (p. 2). Practices range from security purposes to consumer predictions (Pridmore, 2012) to risk management (Amoore & De Goede, 2005) and all sorts of purposes in between. Sorting is rhetorical, then, because it involves classifications of things in particular ways for certain audiences. But back to the key questions: How does a technical communicator work with surveillance information? And how would one begin to identify this surveillance?

Surveillance Workers and Technical Communication

For this conversation, it is most useful to break down the types of information involved in surveillance scenarios that a technical communicator may encounter, and this is best categorized in two ways: (1) information processes by the more formal *Surveillance Worker* like Snowden as scaffolded through the STC's three characteristics of technical communicators (which I argued Snowden was in the last chapter); and (2) the surveillance of information in a more everyday sense through an analysis of various sample essays of surveillance.

TECHNICAL OR SPECIALIZED TOPICS

The first characteristic of technical communication as described by the STC is the communication of technical or specialized topics, thus, technical communicators can be involved in the surveillance process if they communicate technical or specialized topics that are surveillant in nature. While this content could still be communicated *through* technologies, which is the next characteristic that will be discussed, the focus of this point is the technical *content* rather than the technologies information passes through. A little more specific, according to the STC (2018), technical or specialized content might mean "computer applications, medical procedures, or environmental regulations." The topics can have a range, but they all often share similar technical or scientific foundations. Sticking close to the scholarship in the field of TC, more recently, Carradini (2020) looked at TC scholarship to see what topics are important to scholars and noted that popular, current research areas from within the discipline of TC included "knowledge management, professionalization, methods, scientific communication, helping users, China, groups, and pedagogy" (p. 127). More broadly, thinking about the STC's definition, all sorts of specialized topics exist, such as those identified by RAND Corporation (2020) in their "Science and Technology" research. Although they list 26 categories, a sample includes aerospace; cyber and data science; health-care technology; military technology; physics; ships; or science, technology, and innovation policy.

Although not limited to these topics, these provide sample ways of thinking. For instance, topics of privacy and security are certainly in the categories of specialized and technical basis, so it is also easy to see how the conversations overlap. A technical communicator engaged in surveillance work thus would be discussing technical/specialized topics about surveillance. This might be the most recognizable surveillance element because a technical communicator would actively be engaging with themes of surveillance in the content of work, and it would not be hard to identify how the communicator was engaging with surveillance.

Using Snowden as an example of a surveillance here, he would not only have worked with the technological programs, but he would also have worked with the information important to these programs, taken from these technologies that were surveillant in nature (either because they directly addressed surveillance practices or they contained information taken from surveillance). For instance, table 3.1 illustrates a random

Table 3.1. Sample of NSA-Associated Surveillance Disclosed by Snowden

Programs	Description of Program
Bullrun	Program designed to break encryption (Cayford et al., 2015) such as chats or audio of internet chats (Hu, 2015)
Dishfire	SMS text collection (Hu, 2015)
Fairview	Program to collect phone, email, and internet information (Hu, 2015), in conjunction with a partner revealed in 2015—AT&T (Angwin et al., 2015a, 2015b)
MYSTIC DCSN	Collection program (Hu, 2015) for telephone data, to include full content of phone calls (BBC News, 2014)
Pinwale	Storage for email, other text-based content (Ehrenfreund, 2013), and for internet content (Hu, 2015)

Source: Author provided.

sample of five of the 330 programs Hu (2015) identified in her research of Snowden's documents from the June 2013 to January 2015 timeframe, which she noted still "does not purport to be a comprehensive list of all of the Snowden disclosures" (p. 1686). Each example shows how the information is gathered through technologies but also highlights technological content information that would have been a concern in these programs. Bullrun, Dishfire, Fairview, MYSTIC DCSN, or Pinwale are all programs set up to collect data (to subsequently be assessed to become information), and any discussion about their results would have thus involved specialized, technical, and surveillant information.

Again, while the projects detailed above would be technically related, the focus on this example is the specialized content of the programs and how the programs assist in surveillance practices. The projects consist not only of technologies, but also of partner information and additional details of the projects.

Information and Surveillance Technologies

Second, as the STC outlines, another characteristic is that technical communicators communicate information through technology. To be a little more specific, according to the Society for Technical Communication's (2018) work, the technologies being used to communicate information might involve "web pages, help files, or social media sites," but are not limited to those mediums. Technical communicators often communicate through a variety of technologies. For instance, the Library of Congress (2017) notes engineers might communicate through several technical products such as computers or communication systems about their work. As discussed in chapter 1, although in no way a comprehensive representation, scholars writing in publications that deal with computers and writing have noted communication through spaces like codes and algorithms (Beck, E. N., 2015; 2016; Beck, E., 2016; Gallagher, 2017; Vee, 2017), electronic classrooms (Day, 2000), games (Vie, S. & deWinter, 2016), plagiarism detection software (Purdy, 2009), or social media (Vie, 2008). Technical communication publications also talk about games (deWinter, J., & Vie, 2016), discussion boards (Holladay, 2017; Pflugfelder, 2017; Sarat-St. Peter; 2017), comic books (Sánchez, 2020), or information architecture (McCool, 2006).

The idea that technology becomes the vehicle through which information is stored, organized, or passed through is a more basic concept and fits well into discourses of surveillance and information communication devices. This section fits closely with work done in surveillance, because

ICTs are often parts of scenarios of surveillance that are seen to assist in the gathering of information.

A technical communicator then, engaging with ICTs that have surveillance capabilities would thus be engaged in surveillance practices, whether the surveillance work was for a state agency or whether the surveillance work was for a marketing firm gathering information for user experience research. This is especially true in structures of government where Snowden worked. Webster (2012) argues that ICTs "have enabled the computerization of the large databases and record-keeping systems created by public bureaucracies," and along with computers, "telecommunications" have allowed "for the rapid communication of vast amounts of data and information between different locations" (p. 315). The networks and databases also come with the infrastructure of "electronic record keeping, the computerization of large public databases, the ability to communicate electronically and to transfer information/data from the databases to multiple locations using a variety of digital platforms" such as the internet or phones.

Snowden would have technically communicated through these types of technologies. Going back to the sample programs of the last point, shifting a focus from the content of the information Snowden was working with to the technologies he used to obtain the surveillance data provides an example. For instance, *The New York Times* and *ProPublica* reported details about specific technological pieces in the Fairview program such as fiber optics cables, the "use technical jargon specific to AT&T" (which could be both *content through technologies* and technologies themselves), and details about the internet line that serves the United Nations Headquarters, which was subjected to surveillance (Angwin, et al., 2015a; 2015b). In another example, in his own memoir, Snowden (2019a) details communicating to the CIA about using Tor technology for purposes of anonymity (p. 155). Additionally, although Snowden (2019a) is careful to say that he will "refrain from publishing how exactly I went about my own writing—my own copying and encryption" of NSA files "so that the NSA will still be standing tomorrow" (p. 238), he does discuss technologies he used to spy on the spiers, so to speak, when he details removing agency information with an SD card (p. 259).

It is helpful to note, too, that Snowden personally makes the distinction between *content* and *technologies* when discussing the interest in the documents he provided to journalists. He states that journalists "were more interested in the topics of the surveillance reports than in the system that produced them" (p. 222), to which Snowden adds: "I respect that interest, of course, having shared it myself, but my own primary curiosity was still technical in nature. It's all well and good to read a document or to click through the slides of a PowerPoint presentation to find out what a program is *intended* to do, but the better you can understand a program's mechanics, the better you can understand its potential for abuse" (p. 222) (emphasis in original).

Instructions and Surveillance Practices

Finally, a third way that technical communicators can be involved in surveillance activities is through instructions that detail how to do something. As the STC stated, instructions can be "about how to do something, regardless of how technical the task is or even if technology is used to create or distribute that communication," so that leaves the interpretation of what could be considered instructions very open. (However, it is of note, that Sarat-St. Peter [2017] argues that most of the time, the materials are institutional when she states, "Much of the scholarship on instructions focuses on institutional context; this is especially true of technical communication scholarship that predates the widespread adoption of the Internet" (p. 79). Instructions can be a list of the ways to do things, or they can be more formal documents. Kabel et al. (2000) describe instruction manuals as typically being designed "for reference purposes and completeness"; they "generally assume expert users"; they "come in different parts" and often change, requiring updates (p. 2). Rude (1988) adds that they also pay attention to design in their structure and often adhere to design principles that make them easier to follow. For instance, a phone manual may have a list of specifications, list of parts, written instructions, or diagrams centered on being able to work or troubleshoot with the phone.

In a more complimentary sense to points one and two, a technical communicator involved in instructions of surveillance could be communicating about how to operate technologies with surveillant capabilities such as telling a consumer how to set up a smart device; or a communicator involved in surveillance could be writing instructions on what to do with surveillance-related content information taken through surveillance, for example, telling a teacher how to save grade information. However, a communicator may be constrained by rules that restrict movements or what a worker can do with information.

To illustrate again with Snowden, to communicate how to operate technologies of surveillance, backing up to the previous Fairview example, *ProPublica* published PowerPoints from the NSA that Snowden divulged, which show how data needs to flow in the Fairview process. For instance, this document shows how data should flow between corporate, site, and access partners (NSA/CSSM 1-52, 2012, p. 4). This was a point-by-point description of this task.

Regarding what to do with instructions on what to do with surveillance-related content, Snowden himself was more of a working-with-technology person before the disclosures, but especially to carry out disclosures. There are several examples such as when Snowden would "upload tutorials" about how journalists could use encrypted emails to receive content from them (p. 253) because they didn't just want the documents to be dumped to the public; he wanted journalists to sift through the documents (p. 244).

Everyday Surveillance in Technical Communication

Moving on, it is incomplete to explain the connection between surveillance information and the technical communicator solely within a bracket of more formalized roles of Surveillance Workers like Snowden. As stated in the previous chapter, there is a range of what information can be considered surveillance. As also discussed in this book, surveillance can be explained as an everyday occurrence, as well. Thus, that means there are other types of information that can be thought of as surveillance that is not involved in formal surveillance schemes. But to identify what information this covers, one would first need to understand various, less formal surveillance scenarios that exist, and most relevantly for this book, how those would intersect with technical communicators. To explain this point then, I will begin by describing what Marx (2016) might call "surveillance essays" (p. 16), where different scholars try to explain different phenomena of surveillance. Then, to make this information useful, I will look at where surveillance information might crop up in job duties of those engaging in practices of technical communication (more of the lowercase "tc" worker) by providing a tool to help identify and understand surveillance scenarios (which can then be used in preparation of an ethical or social justice analysis, as discussed in the following chapter).

Scholars working in surveillance have described several surveillance scenarios in which one could participate. There are enough varieties of surveillance types that one could even think about surveillance as a Boolean phrase where searching for *veillance (using the asterisk for truncated searching for the word *surveillance*) would be a generative enough search to find various forms of surveillance. Some of the more relevant paradigms of surveillance for surveillance workers are workplace surveillance, consumer surveillance, dataveillance, lateral surveillance, participatory surveillance, and sousveillance.[3] Table 3.2 quickly details the types of surveillance and shows how a technical communicator could be involved.

Table 3.2. Sample of Surveillance Paradigms

Paradigm Description	Communicator's Engagement
Workplace Surveillance	
Workplace surveillance is "the multiplicity of formal and informal practices of monitoring and recording aspects of an individual or groups' behaviour 'at work' for the purposes of judging these as appropriate or inappropriate; as productive or unproductive; as desirable or undesirable; and so forth" (Introna, 2003, p. 210).	Engagement in this form of surveillance can be as either the watcher or the watched for behavior or productivity such as background checks, performance evaluations, monitoring of a workplace internet browser, or watching a coworker's behavior. Workplaces can also entail the observation of other workplaces and monitor for corporate or economic espionage. Marx (2016) even calls the workplace the omniscient organization that measures everything that moves (p. 179).
Consumer Surveillance	
Consumer surveillance relates to the "surveillance of both consumers and their associated consumption" and is reliant on using "personal identifiable data" as a regular source of intelligence for decision making, "corporate practices and products" (Pridmore, 2012, p. 321). This area relates to the previously discussed surveillance capitalism.	Communicators involved in consumer surveillance may be involved in compiling or reporting data and/or information about consumers and their habits to a variety of internal and external partners. The foci in this element are on the business and the consumers.
Dataveillance	
Dataveillance is "is the systematic use of personal data systems in the investigation or monitoring of the actions or communications of one or more persons" (Clarke, 2016).	Those involved in using data through practices like production, human resources, UX, social media monitoring, or any number of data-driven activities can be said to be involved in dataveillance.

Lateral Surveillance	
According to Andrejevic (2005), lateral surveillance is peer-to-peer and "the use of surveillance tools by individuals, rather than by agents of institutions public or private, to keep track of one another, covers (but is not limited to) three main categories: romantic interests, family, and friends or acquaintances" (p. 488).	A person's engagement in this type of surveillance-use tactics is often associated with law enforcement or marketing, but on a more informal, peer-to-peer level, to take "responsibility for one's own security in networked communication environment in which people are not always what they seem" (p. 428). This can range from "casually Googling a new acquaintance to purchasing keystroke monitoring software, surveillance cameras, or even portable lie detectors" (p. 488).
Participatory surveillance	
Participatory surveillance is similar to lateral surveillance in that it is peer-to-peer, but the emphasis isn't on the power relations repeated from, for example, law enforcement contexts but instead focuses on willing participation in surveillance practices in order to engage with others (Albrechtslund, 2008).	Those who engage in participatory surveillance are looking to communicate with others, thus surveillance becomes "empowering, as it is a way to voluntarily engage with other people and construct identities."
Sousveillance	
Sousveillance involves watching those in power from "below" (or those with less power) to challenge surveillance (Mann, et al., 2003).	This can be accomplished when a communicator chooses tactical avenues of delivery to upend power dynamics such as whistleblowing or tactical communication (as will be described further in chapter 5).

Source: Author provided.

In the above examples of surveillance essays, each type of *veillance shows how outside the intelligence community there are a variety of ways a technical communicator who might not formally be known as a Surveillance Worker might engage in surveillance practices. Workplaces, bureaucracies, stores, databases, smart devices, social media, internet sites, a gaze back at those in power—all these examples show the breadth of what it means to be involved in surveillance, and they shows where information used for surveillance purposes can originate. While some spaces may be more conducive to the workplace setting, such as workplace surveillance, others may even engage with technical communicators on a more casual level in their personal time, but as Walls (2017) reminds us, "extraorganizational rhetoric" plays a vital role "in the lives of working professionals" (p. 392).

Now What? Surveillance Scenarios and Putting on One's Surveillance Glasses

I want to further illustrate these surveillance paradigms for technical communicators, but to do this I need to clarify one more thing to maximize usefulness of this framework. To make this information operational, one must "see surveillantly," as Finn (2012) recommends, and notice where surveillance crops up. To make this easier, it is useful to add a brief conversation about methods of identifying spaces of surveillance "in the wild," so to speak, as inspired by Pflugfelder (2017). For lack of a better term, in this section I'll expand on Finn's (2012) idea of seeing surveillantly and describe the identification of surveillance scenarios as "putting on one's surveillance glasses." As briefly addressed in chapter 2, I used the term *surveillance scenario*, because in each instance of surveillance there are at least eight elements that together can form an assemblage[4]: (1) the agent, (2) act, (3) site, (4) target, (5) motivation, (6) information involved, (7) surveillance paradigm, and (8) possible consequences present at any given time (Young & Pridmore, 2021). These elements serve as good entry points into identifying and, as will be explained in the next chapter, evaluating surveillance. The points are somewhat of a mix between Burke's (1945) dramatistic pentad of "what was done (act), when or where it was done (scene), who did it (agent), how he did it (agency), and why (purpose)" (p. xv) and the work of de Graaf (2010), who evaluates instances of whistleblowing (which will be discussed further in chapter 5) and use of the elements "actor, act, outcome, motive, subject, target, and recipient" (p. 76), which can be identified in whistleblowing contexts.[5] For scenarios of surveillance, the agent, act, site, target, motivation, information involved, paradigm, and consequences present can start with the basic assessment question: Does this carry out/limit/resist principles of surveillance (based on your personal definition)? If one thinks the answer might possibly be yes, then it might be worth a more robust assessment of the situation (and then ethics, which will be the focus of the next chapter).

To explain a little about each element, I list each element in the table below, complete with a definition. Each element is critical to understanding a surveillance scenario, because an agent or agents (on possibly many levels) must play some part (act) in a scenario of surveillance located somewhere (the site, which could be various places) and targets someone's or a groups' (target) information for some reason (motivation). Reflectively, it is useful to then assess what surveillance essay/paradigm the agent is operating in (e.g., dataveillance) and then consider potential consequences of the scenario. Table 3.3 provides a brief visual display of these elements.

Table 3.3. Elements in a Surveillance Scenario

Element	Definition
Agent	The agent is the actor who carries out the surveillance.
Act	The act the agent is doing to gather surveillance.
Site	The sites are the location(s) where the surveillance takes place.
Target	Targets are the individual(s) or group(s) under surveillance.
Motivation	This element addresses the purpose of the surveillance.
Information	Information is what is being gathered/assessed by the agent about the target.
Paradigm	This explains the relevant "surveillance essays" in the scenario.
Consequences	The potential consequences of surveillance are explored in this element.

Source: Author provided.

It is also useful to think that there can be various units of assessment for each of these elements, and thus it is helpful to think creatively about who and what is and can be involved. Agents can carry out surveillance on various levels of involvement from primary agents doing the actual surveillance to more tertiary agents who help facilitate the surveillance gathering whether directly or indirectly. Marx (2012) mentions that agents have also been referred to as "watcher/observer/seeker/inspector/auditor/tester" and can be further reduced to roles like "*sponsor, data collector* and *initial* or *secondary user*," p. xxv) (emphasis in original). Acts also can be performed within a range of involvement and may correlate with the list of agents. For instance, in gathering a credit report for a loan applicant, someone (an agent) would design the technology to collect financial information that would then be extracted by someone or something, which would thus be delivered to whomever was then going to be interpreting the information based on other criteria of what is an acceptable risk, and the final act would be the granting or denial of a loan (which might trigger a series of new activities). All along this situation were various actors and acts. In the loan example, the sites could be computers for the data, offices for the assessment, and banks or stores to deliver the results of the assessment to the credit holder. The targets[6] might be the credit applicant or possibly anyone at their level of risk that they could be compared to.[7] Motivations for the surveillance might be initiated by the target to get a loan, and the information gathered would be financial records. The paradigm of surveillance could be credit surveillance as part of dataveillance, and the consequences could be that someone does or does not get a loan, which could lead to workplace surveillance, say, if the credit report was needed for some type of personal business loan. Consequences could be the loss or gain of a loan and other tangentially related to motivations and consequences from the gathering, sorting, and acting on the surveillance information.

Additionally, often it is also useful in certain situations to note applicable laws or regulations covering the scenario or historical context, and maybe even make a specific point to assess the technology involved. Marx (2012) also emphasizes the distinctions between organizational and nonorganizational surveillance and internal and external classifications (p. xxv). While these elements are all useful and these and other classifications underscore a range of important details that could be included, the ones I have laid out here are useful for the basic who, what, where, when, why, and how questions that help journalists and investigators get a general handle on a situation.

Operationalizing "Surveillance Glasses" through TC Analysis of TC Information

It is now useful to further operationalize the conversation of surveillance paradigms while "putting on one's surveillance glasses." There are four frequent ways a technical communicator can participate in surveillance processes using the surveillance glasses method while working with information not readily identifiable to Surveillance with a capital S.

To begin with, in TC literature there are recurring locations where communicators either get information, transform data to information, or use information. Four of these spaces are (1) documents, (2) content management, (3) information architecture, and (4) user experience (UX)/usability. Breaking each of these down into their role in processing information is beneficial for the overall purpose of this section.

First, documents are what Lauer and Brumberger (2016) might call "information products" and are central to TC and the organization delivery of information. According to them, "Developing information products . . . has long been the core of technical communication work" (p. 249). Deliverables of information can range from the more conceptualized document such as a handwritten or printed, physical file, or electronic product, including emails, memos, letters, reports, brochures, technical manuals, user manuals, or instructions for at-home building. Documents can also be categorized in various ways such as the informative document or the instructional, persuasive, or corresponding document (Kmiec & Longo, 2017), but even when a document is not officially thought of as "informational," instructions, persuasion, and correspondence all must deal with transmitting information to an audience. Documents are an ideal place for information to be sorted and organized for usefulness, because they are such visible conduits from communicator to audience. For instance, Kmiec and Longo (2017) note, "[R]eports are a key way work is communicated" (p. 94). If a technical communicator wanted to relay information to someone, creating an informational document would be a key place to do that, as it can be arranged, organized, and transported from one place to another. In a surveillance scenario, a document is a very broad concept to engage with, but table 3.4 illustrates one concrete example with a performance review report. In this example, it is helpful to imagine an employee entering information into computer software about their work performance that can then be downloaded into a report by their supervisor for the purpose of an annual review.

While for a specific situation, any of these variables could be explored in greater depth, and one could argue for or against additional variables, they at least provide a scaffolding to reveal how one could be involved in a chain of surveillance scenarios. (This could be the same for the rest of the examples, too, and should be kept in mind for the remainder of this section.)

Table 3.4. Analysis of Performance Review Report

Element	Example 1: Performance Review Report
Agent	Supervisors at any level tasked with assessing an employee's performance; employees are also agents if they are required to compile their own evidence. The computer program design also plays a role in limiting the information that can be complied.
Act	Performance review.
Site	Workplace, department, computer, software, various offices, meeting location.
Target	Whoever is accountable for the review, first being the employee and then the various levels of management tasked with maintaining productivity.
Motivation	Performance review for continued personal and institutional effectiveness.
Information	Productivity, output, evidence of service, record of infractions, absences, trainings, list of responsibilities, past/present/future goals, etc.
Paradigm	Workplace.
Consequences	Promotion, status quo, job loss for any of the targets involved. Reevaluation of expectations, new responsibilities, changes (good/bad) in attitude, etc.

Source: Author provided.

Second, on a basic level, content management helps organize and manage information in digital spaces, and it can be focused on a variety of things, be it managing content on the web or managing content for an enterprise (Batova & Andersen, 2017; Andersen, 2007). Content management is a particularly rhetorical and surveillant activity because it involves the selection of information for delivery to specific audiences. Hart-Davidson et al. (2007) illustrate the rhetorical nature of information in content management when they argue that CM "is a deeply rhetorical process for organizations to undertake" (p. 14), because the selection of information of what to include on, for example, a website, such as why an organization exists or its mission statement, is techne and an argument in itself, especially for content on public-facing systems (p. 11). Table 3.5 illustrates this area with the example of content managed for a professional, self-created web page.

Table 3.5. Analysis of a Personal, Professional Web Page

Element	Example: Personal Web Page
Agent	Self-created page with the help of free, online technology.
Act	Creation of a personal website.
Site	Online space.
Target	Personal highlights of important information.
Motivation	Increased visibility of one's work and more business prospects.
Information	Content selected supports the motivation of increased visibility but is constrained by the affordances of the software and hosting site allowances.
Paradigm	Participatory surveillance by willingly providing information for visibility. Workplace surveillance allowing others insight to one's work. Dataveillance.
Consequences	Contact information is readily available, personal details on the web and stored in someone else's computer. Possibly permanent record of activities.

Source: Author provided.

Further, it is still of note that one can both be an agent and target of surveillance, such as in this case, and as soon as one starts to look at someone else's page, they can also move into a more prominent agent position.

Third, as described by scholars, information architecture is "the presentation of information in meaningful patterns" (McCool, 2006, p. 169). Crystal (2007) describes the activity through four components: organization, navigation, labeling, and searching, and these activities entail that information architects must organize information in "a hierarchical structure that is comprehensible to users," as well as help users move through these spaces by creating and labeling categories that make sense and can be readily and easily searched (p. 16). Information architecture is an "information product," because at the end of the organization, there is a tangible deliverable that organizes and labels information that allows navigation and searching in certain ways. Table 3.6 illustrates these connections, looking at a more formal employment site made by a vendor.

Table 3.6. Analysis of Institutional Website

Element	Example
Agent	Third-party vendor, targeted institution providing details about itself, technology that constrains what can be shared. Audiences watching a corporation.
Act	Creation and maintenance of an institutional website. Organization plays a role in making the content either visible in the first place or more visible.
Site	Online, institutional (technology provider, vendor, institution itself).
Target	Institution and employees at the institution.
Motivation	Website for visibility and showcase of the institution such as its ethos, employees, and information.
Information	Audience-specific institutional capabilities, strengths, previous work, contact information of business and employees.
Paradigm	Workplace surveillance, sousveillance, lateral surveillance.
Consequences	Increased business and revenue, unmet expectations, (unwanted visibility) of employees, ideological (dis)agreements, etc.

Source: Author provided.

In this case, selected paradigms were workplace surveillance, sousveillance, and lateral surveillance, which helps one think of which variable to emphasize; however, one can't start from the paradigms without already sketching out the other elements. It helps to work back and forth. For the paradigms selected, workplace surveillance can be more about workplaces watching other workplaces for purposes of competition or possibly even espionage. Sousveillance could come in the form of an environmentally conscious audience watching a corporation like an oil company detailing its future production goals. Lateral surveillance could come from coworkers or others looking at personal profiles to assess credentials of peers or conduct informal, online background checks on personal hires based on résumés on file. Its power balance is more equalized. Overall, though something like an employee profile page with faces, dates of employment, and office numbers, may not be readily "surveillant," all of its elements could allow someone to find and watch someone else. It would be interesting to know if a company would allow employees to opt out of personal profiles.

Finally, a fourth way TC discusses rhetorically organizing information is through the concepts of usability and user experience. This area of research is devoted to analyzing the audiences' abilities to access and experience information, and this is important because as Opel and Rhodes (2018) point out, messages don't end at the creation of it; rather, communication lives on in the users of it (p. 71). Although Haaksma et al. (2018) report that usability is a complicated concept that encompasses a large range of characteristics like efficacy, efficiency, or learnability, "and consensus has yet to be reached" on the diverging ways these concepts have been referred to (p. 118), Lauer and Brumberger (2016) provide at least a baseline definition that "[u]sability focuses on evaluating how well a user can navigate through a variety of tasks that an end product was designed to facilitate" (p. 249). Haaksma et al. (2018) describe UX as a broader concept and a response to usability. Some criticize usability for putting too much focus on task efficiency, work, and instrumental factors, and instead, UX looks at "the interaction between people and products, and all sorts of experiences that result from it" (p. 118). As Rose et al. (2017) summarize, UX involves implementing usability studies to see how a "product meets the needs of users," and it involves understanding one's sensory responses to products, systems, or services and "the process of designing software and systems and the outcomes of the interactions with those systems" (p. 215). Lauer and Brumberger (2016) add, "UX also strives to accommodate how users appropriate information products and content in unanticipated ways and for their own purposes as well as how those products position users to act in the world by the way they are designed and the options they allow for" (p. 249). Table 3.7 illustrates these connections to surveillance with a social media user experience. It is helpful to imagine a firm conducting a study involving social media statistical data and a focus group of users and their response to a social media site.

An example of UX for social media pages provides a robust example of surveillance practices involved in everyday aspects of TC in the form of big data from social media use, dataveillance, and even surveys of one's experiences of interacting with sites of participatory surveillance. The UX research can be used for the purpose of consumer surveillance to possibly adapt and deliver a better social media products for more profit. Agents are people who collect or look at data. The act involves analyzing one's experience with a social media site, which can be where the actors overlap with the locations of the targets. Targets are those who initially post the content and those who review their experiences of that content. Consequences can range from being positive for the company, for exam-

Table 3.7. Social Media User Experience Analysis

Element	Definition
Agent	UX researcher.
Act	Analysis of social media UX through big data, data analysis and surveys.
Site	Online, institutional (UX worker), additional locations (site of creation of posts or site of responses to UX research).
Target	Social media user and original social media content producer.
Motivation	More effective social media product.
Information	Details about social media use, social media site information, personal information, metadata, etc.
Paradigm	Consumer, dataveillance, participatory surveillance.
Consequences	More revenue, mass of data that needs to be secured, potential information leaks, (in)appropriate sorting of information, biased algorithms, etc.

Source: Author provided.

ple, when people bring in more revenue, to being more problematic, for example, when there is a resultant increased need for data storage and privacy policies (which would be a good thing for those who are targeted), or when questionable sorting results are either inaccurate or troublesome because algorithms are biased.

Overall, even when people do not consider themselves to be "surveillance workers," they can be participants in the larger surveillance scenario involving the processing, manipulation, or organization of non-"traditional" surveillance content obtained or practiced in one of the surveillance paradigms mentioned above, such as workplace surveillance or dataveillance, or especially in the case of UX, the gathering of others' information through which one can make data-driven decisions on how to manage products and people in the future.

Conclusion

This chapter illustrated that TC has a varied range in terms of how it deals with information, which is often the basis for surveillance work, but it also shows that these exist outside the confines of strictly state or hierarchical scenarios. Tasks of surveillance can come from both the more formal activities of surveillance such as those of Snowden to activities more characteristic of everyday processes, where technical communicators not involved in the intelligence community might operate. The following chapters now begin to address what one can do with the knowledge of potential involvement in surveillance paradigms, whether it be to ethically evaluate one's role, resist it, or explain it to others.

Chapter Four

Evaluations and Responses
Social Justice, Ethics, and Surveillance

Building on the last two chapters, this chapter takes up the conversation by answering, so now what? If chapter 2 argued that there is a range of technical communicators, and chapter 3 added that this also involves a range of what can be considered surveillance information, then this chapter approaches the topic of why we'd even want to focus on surveillance in the first place. So what if there is surveillance? What does it matter if someone is watching? The oft-repeated phrase of the "nothing to hide argument," something akin to, "Why worry about surveillance if you're not guilty of something?" is a popular refrain circulating in the popular imagination (Solove, 2011). But surveillance can and does matter, even if you have nothing to hide. It always has consequences. And those consequences are experienced and felt differentially. That is why multiple approaches to evaluating the potential consequences are necessary when considering surveillance, such as approaches of social justice or ethics.

Consequences, some of which we've already talked about, can come from the datafication of society and the increasing use of data as a synecdoche for real people as data doubles (Haggerty & Ericson, 2000), to biased algorithms (Dencik et al., 2017), to various forms of abuse (Fauci & Goodman, 2019). There can also be diminished trust, continued or increased imbalances of power, errors and false positives in system classifications, and social sorting, which discriminates against certain groups (Macnish, 2018). There are also risks to creativity and the reduction of taking risks (Richards, 2013; Zhang & Saari Kitalong, 2015), to the chilling of civil liberties (Richards, 2013), which happens "when people choose not to engage in legitimate activities for fear of being monitored" (Macnish, 2018, p. 35). Penney (2017; 2021) and Shaw (2017) illustrate that people change their internet search habits to self-censor when they think they're being watched.

Summarizing many of those consequences, one could argue that surveillance is *oppressive*, be it the disproportional targeting of certain groups, the unfair classification or treatment of people or groups, or a chilling effect that one has lost a sense of freedom or exploration. In this take on surveillance, TC scholarship is especially poised to provide a rich discussion of how surveillance oppresses and how we could react to it through the field's treatment of social justice, as illustrated through the work of Walton et al. (2019), who not only work through what oppression looks like but also offer strategies to address spaces of oppression.

It is of note, however, that while viewing surveillance through consequences such as oppression is useful for arguing its importance and taking action against the injustices that its effects can cause, it is limited to talk about surveillance through the rhetoric of "harm." "Harms" not only conjure up an image of bodies in the street, so to speak, of which surveillance "harms" often don't lead to (Solove, 2011), but it also places an overwhelming emphasis on the negatives of surveillance that can both ignore the benefits of surveillance and alienate those that from the onset don't preface surveillance in terms of "harm." As Lyon (1994) and Wise (2016) argue, surveillance has two faces. Some of the benefits derived from surveillance for both individuals and corporations are safety and security (Lyon, 2007), social connectivity (Albrechtslund, 2008), productivity (Harwell, 2020), and income (Andrejevic, 2002; Fuchs, 2014). Surveillance can make life easier for things like shopping recommendations, and the internet of things (IoT) can make life easier when "things" get hooked up to the internet and can transmit data. I can know at the grocery store when I need milk at home, even if I didn't check before I left. Some even get pleasure from watching others (Wise, 2016), and one of the foundations of surveillance in general is a rhetoric of care (Wise, 2016, p. 11). As will be illustrated in this chapter, national security can be argued as a positive consequence often framed around concern for safety.[1]

Another strategy to go about looking at situations of surveillance for their consequences then, is to use ethics as a heuristic. Situating events through ethics helps identify why certain behaviors might be acceptable or problematic depending on specific circumstances and from variable viewpoints (Kupperman, 1969). While ethics as a heuristic is somewhat fashionable in postmodernism (Porter, 1993), it has also been criticized for its focus on the individual (Walwema, Colton, & Holmes, 2022). However, when used in addition to frames of social justice, both approaches complement each other.

With that premise in mind, this chapter argues that while social justice research can help us see that surveillance is oppressive, ethics as a heuristic can also help see why someone would still support systems of surveillance. Both perspectives speak to how to talk about it, and how to react to it, especially from the premise that not everyone will agree on its consequences.

Ethics, Social Justice, and TC

To come to this conclusion, it's useful to step back and see how ethics and social justice have been situated in TC, often as less than complementary frameworks. There are various approaches to ethics in TC scholarship topic ranging from (but in no way limited to) explorations of cultural communication and philosophy and ethics (Dragga, 1999; Markel, 2001), workplace ethics (Dragga, 1997; Faber, 2001; Markel, 2001; 2005), design ethics (Salvo, 2001), ethical literacies (Cargile Cook, 2002), and pedagogies and textbook praxes teaching ethics to students (Allen & Voss, 1997; Dombrowski, 2000; Markel, 2001; Selfe, 2007). Other more recent topics argue a need for more work on ethics in general (Walwema, Colton, & Holmes; 2022); ethics and invention (Boedy, 2017); institutional review boards (Phelps, 2022); ethics at borders (Pihlaja, 2022); disability rights, justice, and access (Bennett & Hannah, 2022); teaching ethics in TC (Lee, 2022); and ethics in editing apps (Wang & Gu, 2022).

From a more pragmatic view of the workplace context, Fischer (2004) adds that business ethics can be understood as the evaluation of "the rules, standards and moral principles" of personal and organizational choices in a business setting. Powell (2019) also describes how a technical communicator in a workplace context can use ethics to communicate clearly and effectively to describe objects, processes, and procedures in morally acceptable ways. Steps to do this include making sure to relay information clearly and accurately, using appropriate language, adhering to copyright law, dutifully analyzing data, minimizing bias in communication, and overall making thoughtful decisions on anything that needs a well-rounded, consideration for various stakeholders' interests.

It's important to note, though, that while ethics are often framed as individual endeavors (Jones et al., 2016), ethics also require attention to the larger context. Dombrowski (2007) illustrates this when he describes how trends in TC research consider ethics not just as individual, personal matters, but also as matters of "the complex social context in which individuals think and act" (p. 307). He goes on to emphasize that while some (particularly older texts) stress the importance of the individual, the social setting in which that individual operates is an important space to consider, for example, in the context of international communication and sensitivity to varying cultures (p. 316).

In terms of language and concept, moving a conversation of ethics toward the more complex social context that focuses on community often results in shifting the conversation from "ethics" to "social justice." This shift is reflected in Jones et al.'s (2016), comment summarizing that the "social justice turn" in TC involves "moving from mere ethics, which often exist in an individual's character or behavior, to a social justice stance, which tends to be more collective and action oriented" (p. 211). While the addition of "mere" in front of ethics might downplay its weight and importance, and while Walwema, Colton, and Holmes (2022) critique ethics as being male and European and American-centric, Walwema, Colton, and Holmes (2022) also remind scholars that social justice and ethics are inextricably connected, as "social justice undeniably stems from moral values" (p. 258).

Taking Ethical and Socially Just Action

One particularly useful benefit of both social justice and ethics as frames is that they not only help us to analyze situations, but they also help us react to them. While their reaction paths differ, both direct those who have identified spaces of ethical murkiness or social injustice to act. The ability to tap into schemas for action further emphasizes why a good background in ethics and social justice would be useful, because one would need an ethical or just literacy[2] in the first place to recognize (un)ethical or (un)just spaces to react to.

To explain ethical action, it is helpful to turn to the work of Manuel G. Velasquez (2012) in his book *Business Ethics: Concepts and Cases*, where he adapts the work of moral psychologist James Rest (1986) to the business context. Drawing from Rest, Velasquez argues that there are four processes involved in acting ethically. These steps are: (1) recognizing an ethical situation; (2) making a judgment about what an ethical action is; (3) deciding what the right thing to do is; and (4) carrying out the decision (pp. 49–56).

To explain each a little more, to act ethically, one must first recognize an ethical situation in the first place, and that involves using two subframes of assessment. The first subframe requires recognizing the rhetorical situation of the space. This involves assessing stakeholders, messages, and evidence in the given situation and evaluating how an audience might interpret them. Talking to one's boss at work may be different from talking to one's toddler at home. The second subframe involves assessing *to what degree* an action will meet the following six criteria: (1) it will cause harm; (2) it will cause a significant amount of concentrated harm; (3) the harm is likely; (4) any victims are proximate to us; (5) the harm is imminent, and (6) the harm is a violation of our morals (p. 50). Both subframes require a situational assessment.

The second step to taking ethical action is judgment of what is an ethical choice is in the first place. To do this, first, it would be necessary to grapple with one's own personal biases about (1) the world, (2) others, and (3) oneself. We'd need to think about how we often view the world though a narrow framework. We'd also want to think about how we position ourselves in relation to others, reflecting on stereotypes. Finally, we'd want to evaluate ourselves and how we might have unrealistic ideas about ourselves, that "we are more capable, insightful, courteous, honest, ethical, and fair than others." We are also often "overly optimistic about our futures" and "overly confident about what we think we know" (pp. 52–54).

Third, to engage in ethical behavior, one would need to determine the right course of action to take and commit to taking the ethical path. Unethical people don't always start out that way. Ethical people can be enticed, persuaded, or feel trapped into making unethical decisions based on their surroundings. For instance, over forty years ago, Parker (1976) commented that computer crime offenders "tend to be very ordinary, [and] they simply had a problem to solve or somehow developed a goal to achieve beyond ordinary means. They were in a position of trust that could achieve their purposes" (p. 43). Some corporate/government/institutional cultures can be toxic and place employees in questionable circumstances, which can (in)directly encourage their employees to make bad choices to be "team players," or fall victim to "moral seduction" where one engages in slightly more problematic behaviors until they are fully practicing unethical actions (Velasquez, 2012, p. 55). People who make ethical choices may have to make that choice even to their own detriment, but an ethical action involves determining what is ethical and then committing to that position.

Actually doing what is right is the final step of ethical action. This requires that we not only regulate ourselves so we do what's right, but also that we think about the external factors we might encounter, whether they are matters of luck or a result of the power of others. The idea that power impacts one's ability to carry out ethical decisions is especially important because "many people willingly obey authority figures even when they believe or suspect they are doing something wrong" (p 56). Because surveillance is especially tied to (often asymmetrical) relationships of power (Fuchs, 2011), it is especially important for those in situations of surveillance to be able to do what is right and stand up against others with power. "I was only doing my job" has been a problematic refrain throughout history, for example, when lower-level Nazi's carried out the heinous orders of their superiors (Barajas, 2016).

A path based on principles of social justice follows similar steps as ethical action, albeit with differently placed emphases. Like ethics, one using a social justice lens would first want to evaluate the rhetorical situation. Walton et al. (2019) note the need for "critical context analysis" (p. 143). However, questions of social justice would likely diverge from the six ethical queries noted for this assessment, disassociating from inquiries that sound dubiously similar to Yes, this will lead to harm, but how *much* harm? in favor of questions such as What forms of oppression show up in this action? Who does this activity oppress and why? How would my actions lead to more oppression or more justice?

A second step to determine a socially just action could utilize Walton et al.'s (2019) 3Ps—positionality, privilege, and power—to grapple with personal biases. As noted briefly in chapter 1, positionality encourages the examination of often contradictory positions one is coming from. Privilege reflects on unearned advantages we may have been given; and power flows through structural, disciplinary, hegemonic, and interpersonal communication (p. 115). Positionality is useful because it "contextualizes people in complex ways that allow for tension, for conflicting truths, for imperfection, for things that seem like they don't fit or can't coexist to do so" (p. 80). Interrogations of privilege are useful for thinking about which groups of people are the most "easily believed, valued, unquestioned, protected, and unburdened by the responsibility of who they are and how they are" (Walton et al., 2019, p. 102). A critique of power is important because the intertwining dimensions of power illustrate how social dynamics influence discourses and the way that people are (indirectly) influenced to live their lives and interact with each other. Surveillance is particularly a matter of position, power, and privilege, because, as is repeatedly emphasized, who gets to watch, who is the target, and how the watching is leveraged and used is often a question of power—more specifically, who doesn't have it and who does.

Third, like ethics, social justice also motivates actions to determine the right course of action to take. Like ethics' third step, which requires thinking about what an ethical action is and committing to that path, one concerned with social justice would also want to understand what is a socially just action is and commit to that action. In the Snowden context, Snowden could be said to have determined it was ethically wrong to engage in mass surveillance and committed to sharing with others what he knew.

Finally, social justice–driven actions would also lead up to taking actions, just like ethical decisions, but ones that are ostensibly more collective. One defining hallmark of social justice, especially in comparison to ethical decision making, is that it is "coalitional action in support of justice" (Walton et al., 2019, p. 63). As noted, oppression is structural and institutional, and it takes many to make change. One cannot address systemic exploitation, marginalization, powerlessness, cultural imperialism, and violence alone and expect to get results. Actions taken through a lens of social justice then would also most likely need or lead to more collective decision making. That takes not only participation but also the assurances that others are allowed to speak, even if that means staying silent oneself. For instance, as argued by a panelist during the Data Justice 2021 conference (Data Justice Lab, 2021), we need to step back to hear the Muslim woman speaking about her experience.

In a parallel conversation to ethical decision making in general, Walton et al. (2019) offer another useful heuristic that also echoes the ethical-decision-making framework when they advocate for the 4Rs: recognition, revealing, rejecting, and replacing. These steps follow this path:

- First, one must *recognize* injustice and oppression before doing anything with it. One can't act on oppressive practices unless they recognize them in the first place (p. 138).

- Second, one must *reveal* injustice and oppression (p. 139). Revealing is a call to action to point out to others unjust and oppressive practices. Such as in the examples of surveillance, revealing oppression could be done through whistleblowing practices, posting online videos, or it could take more of a guerrilla form, where one speaks out to others about standing up to an institution or even a friend who participates in forms of oppression through surveillance.

- Third, after one reveals the injustice and oppression, one must also *reject* it through the refusal "to support the behaviors and structures that oppress groups of people and leave them at the margins" (p. 141). This can range from rejection through dialogue and then lead to more action through mutual agreement and then action to correct the injustice.

- Finally, fourth, the *replacement* of injustice and oppression involves changing and substituting an "oppressive behavior, structure, or decision" with something else, whether it be personal change, modification of a procedure, or a replacement of structural practice (p. 143).

These four steps offer yet another way to think about not just evaluating spots of surveillance but actively reacting to spaces if they are determined to be consequential.

Overall for this section, while neither ethics nor social justice would be constrained by either paths or motions, both social justice and ethical assessments do support particular ways of movement, and both emphasize two distinct routes: one more individual and the other more collective. I will pick up on these types of movement after a deeper look at surveillance and Snowden.

Social Justice, Oppression, and Surveillance

Back to the larger argument, surveillance matters because of its consequences, in particular, its ability to oppress. Equating surveillance with oppression is a meaningful move because no matter how good surveillance may be for someone, it will likely result in some degree of oppression for someone else. While surveillance may help with product recommendations or competitive insurance rates, it can also alter the lives of those it touches by creating categories of marginalized consumers or by making it more expensive for one to acquire goods—services like insurance, or basic necessities (Loi et al., 2020).

TC scholarship is particularly poised to extrapolate the *oppressive* nature of surveillance through the field's treatment of social justice. Social justice and oppression are inextricably linked because as Walton et al. (2019) comment, "[o]ppression makes social justice necessary" (p. 17). When we are talking about oppression then, we are also talking about social justice. Walton et al. (2019) outline nicely what oppression means, drawing from a variety of scholars but in particular the work of Iris Young (1990) and her classification of the five faces of oppression: exploitation, marginalization, powerlessness, cultural imperialism, and violence. These oppressions can appear singularly, or they can be intersectional where "systems of oppression are interlocking, overlapping, and experienced all at once by those who are multiply marginalized" (Walton et al., 2019, p. 28). Walton et al. summarize these five faces in the following ways.

First, oppression is *exploitive*, when people don't benefit fairly from the work that they do. This can come from overwork and underpayment, and it involves dehumanizing and devaluing the work that some workers do through justification that somehow the worker doesn't qualify for equal compensation.

Second, oppression *marginalizes* by placing people and groups to "the edges of societal and organizational decisions, cultural representation, and legitimated experience and expertise" (p. 19). Those at the margins are outside of the typical "norms," whose needs are not given the same level of attention as those more centered in society. Typically, in places like the United States, the norm defaults to "white, straight, male, middle class, and Christian," as well as "abled-bodied and cisgender" (p. 19).

Third, oppression creates *powerlessness* through a lack of autonomy, authority, creativity, decision making, and respect. When people are powerless, they do not have the command to make decisions and are not listened to because their ethos and role identities are not recognized as valuable. Conversely, those who are used to being valued are often "oblivious to the advantages of automatically commanding respect in their knowledge, authority, and perspectives," but those without that power often feel helpless (p. 24).

Fourth, *cultural imperialism* is a form of oppression that "makes invisible the perspectives of oppressed groups," while also stereotyping them, making them the Other, and erasing their experiences and culture (p. 21). This form of oppression again places a dominant group in the center as a norm "by which other cultures are judged," and often categorizes people by how "white" they are, as well as fails to recognize the knowledge and value of those who are Othered (p. 22).

Finally, fifth, *violence* oppresses and "includes attacks and threats of attacks, physical and psychological violence, harm to people's bodies and harm to their possessions," as well as reliving those experiences through platforms like mass media (p. 24). Some forms of violence are overt, for example, rape or murder; other forms "aim to degrade, humiliate, or destroy" and can be more subtle, for example, microaggression or blaming oppressed people for their own oppression (pp. 25–26).

Making the comparison to surveillance, these five faces show that surveillance can be unquestionably oppressive, and very often intersectional. First, surveillance is *exploitive*. While in no way is this an exhaustive list, exploitation particularly shows up in workplace monitoring software where workers are monitored for production (Andrejevic, 2002), such as work-from-home technologies (Harwell, 2020) or warehouse fulfillment metrics. The near nonstop production measurement and constant visibility of the workers' bodies can benefit the bottom line of an employer while not benefiting workers, and sacrificing their health (Bose, 2020). It is also a system that uses the unpaid labor of social media users (Andrejevic, 2002; Fuchs, 2014). Or it takes the data from anyone in its web, often unwelcomed or without permission, and uses that data for its own end, often without compensation, sometimes in parasitic ways, and to the detriment of its host.

Second, surveillance *marginalizes* those under its control, often through its function of organization. Gilliom (2001) comments that surveillance systems, are not just ways of watching but also ways of seeing, that "impose an order upon the world that shapes both our understandings of reality and our capacities for action" through decisions about what does and doesn't matter, and how rewards, benefits, punishments, and costs should be distributed (p. 9). One process that illustrates this is through social sorting (as discussed in chapter 3), which Lyon (2009) describes as "the use of searchable databases and associated techniques such as data mining, characterized by the classifying and profiling of groups in order to provide different levels of treatment, conditions or service to groups that have thus been distinguished from one another" (p. 41). In the case of consumer surveillance, for instance, Gandy (2011) argues that a technology that transforms "transactional-generated information" into "actionable intelligence" is a form of "statistical discrimination," because its "sophisticated analytics contributes to the cumulative disadvantage that weighs down, isolates, excludes, and ultimately widens the gaps between those at the top, and nearly everyone else" (pp. 175–176). Lyon et al. (2012) comment that this sorting is a form of institutional processing and can result in "vastly different life chances" (p. 3). Five years prior, Lyon (2007) had commented, "All too often, it is the already-existing categories of 'race,' nationality, gender, socio-economic status, or deviance that inform and are amplified by surveillance, which then enable differential treatment to be given to the "different" groups (p. 183). Surveillance becomes an epistemology of truth or legitimization that argues some categories are the "ins" and some are the "outs." Some zip codes get marked as a higher risk by servicers while others receive less attention or fewer perks (Weinberg, 2017).

Third, surveillance also creates *powerlessness*. For instance, the whole notion of panoptic power and Foucauldian discipline is built on the idea that watching curtails the watched person's ability to deviate from the ways that the watchers want the watched to act. Further, John Gilliom (2001) illustrates other forms of powerlessness through his work looking at how those on welfare are watched. He references that those watched by case workers and the state often feel that their relationship with the state is one-way, with only the agency having authority (p. 162), and some feel that even legal rights are "irrelevant, unreachable, or remote" due to the powerlessness of their position (p. 11). In addition, Snowden argued that terrorism rhetoric, especially without any specific threat, is an attempt to make the American public feel powerless, and that the political class has made "a cynical attempt to turn terror into a permanent danger that require[s] permanent vigilance enforced by unquestionable authority" (Snowden, 2019a, p. 205).

Fourth, surveillance also facilitates *cultural imperialism*, especially through its routine monitoring of borders (Adey, 2012) and its fundamental classification of which citizens are "in" and "out," leading to different life experiences. As Lyon (2009) notes, the state uses "counting, monitoring, and documenting citizens" as a form of power (p. 4), and these identifications have evolved to include "not only legal and political rights but economic and social ones as well" (p. 24). These state-run identification systems have historically served as "an ideal means of distinguishing between different sectors of the population and meting out different—

not to mention highly unjust—treatment" (p. 27). Surveillance has also been used more informally where citizens watch cultures different from their own to single them out for mistreatment online (Plesnicar & Sarf, 2020). Themes of cultural imperialism are particularly present in national security rhetoric because national security often prioritizes the "national" at the expense of delegitimizing the Other. This is particularly seen in post-9/11 rhetoric where "us" versus "them" often shows up. Although connected to questions of constitutionality as well, that surveillance data was maintained on American citizens was marked by rhetoric as being something especially egregious (Toomey & Gorski, 2021), yet is somehow rhetorically more acceptable for certain groups on the "outside." The surveillance concept of the "banopticon" highlights the post-9/11 push to systematically exclude certain people on the outside (Bigo, 2006).

Finally, surveillance can be a form and tool of *violence*, both causing physical or mental harm. Physically, surveillance can be executed through force or coercion, as in the case of national and citizen surveillance of the Jewish population during World War II. The idea of the "militarization" of surveillance involves potentially violent, militarizing practices of watching in civilian life (Ball & Snider, 2013). Mentally, the threat of violence or force or the perpetual humming of a drone (or the fear of its invisible presence) can affect one's well-being (Bauman & Lyon, 2013) and turn someone's backyard into a space of state visibility (Jensen, 2016). Surveillance can be harmful such as in the everyday sense when social media users watch each other. Reports show how teenage girls are especially bothered by the continued use of Instagram and often compare their lives to the seemingly "perfect lives" of others (Wells et al., 2021). Chilling effects can be a form of violence if one becomes fearful enough of surveillance practices as to inflict mental harm or fear (Penney, 2017). Snowden himself talks about the mental consequences of surveillance for both himself and one of his targets when he comments that the one time he met a target in person was "unforgettably visceral and sad" and that the target went on to be arrested, lose his job, and was forced from his home through the set of motions Snowden set up even though the man refused to work with the CIA (p. 157).

All of these ways surveillance oppresses is thus very important to look at because we can see surveillance's consequences are oppressive and do not contribute to a more just and equitable world. But again, while a framing of oppression is especially useful for surveillance because it helps us see how surveillance is consequential in that it causes social injustice, that framing is also limited because there are also benefits to surveillance as well, and often the harms and benefits can be felt at the same time by the same people.

Ethics, Snowden, and Surveillance

A complementary framework to the idea of oppression is ethics, or in simple terms, a study of right, wrong, and moral (Velasquez, 2012). Ethics can serve as heuristics and entry points through which to assess and examine a specific situation to identify spaces of conflict, but it also starts with a more philosophic exploration of what is good and bad (Kupperman, 1969), which helps to avoid starting out with the premise that surveillance is oppressive. This also reflects the discursive approach to surveillance, which surveillance ethics scholar Eric Stoddart (2012) argues means, "ethics are not the outcome but the process itself" (p. 373).

While an examination of why surveillance would not be oppressive could also be assessed, this book will use ethics to illustrate why not everyone might agree that surveillance is oppressive. The points of disagreement are critical to look at, because often both positions could come from a place where each side may wholeheartedly believe they are taking the "right" action. As anecdotal support, after working in the background investigation industry with coworkers with a variety of political leanings, I found that they justified their beliefs from positions of care that reflected wildly divergent opinions on topics such as immigration. Some positioned their locus of care to those immigrating, and others centered their care on their more immediate families and saw immigration as something threatening to those they cared for. Both called on the rhetoric and ethic of care, and this is why both sides also benefited from deliberation, consensus, and dissensus and making more visible diverging viewpoints (Knievel, 2008). Ethics help draw out these various viewpoints, and in particular, this chapter draws on the work of Velasquez (2012) to outline four areas of moral ethics: utilitarianism, rights, justice, and care. A brief explanation of each of the four areas is useful before continuing.

Utilitarianism

The first category of ethical decision making that Velasquez (2012) outlines is *utilitarianism*. The word *utility* refers to the overall "benefits produced by an action" (p. 78). Building from that, *utilitarianism* is a moral, ethical view that "holds that actions and policies should be evaluated on the basis of the benefits and costs they will impose on society." Benefits could mean "pleasures, health, lives, satisfactions, knowledge, [and] happiness," and harms can include things like pain, "sickness, death, dissatisfaction, ignorance, [and] unhappiness" (p. 78). It is a general outlook that asks the decision makers to weigh their decisions based on what will produce the most desirable outcome for society. Right at the start we can see the benefits of ethics as an analytical frame by not only focusing on the costs of something but also the benefits.

Velasquez provides the example of Ford Motor Company and its 20th-century vehicle design for the Pinto as an example of the utilitarian approach to decision making. In testing, the company found that the gas tanks had the potential to break apart in some car crashes. However, instead of altering the design, the company decided that what was best for society was to build the model anyway, to make a less expensive car with the potentially explosive capabilities. They did this because the vehicle did meet safety standards, had overall safety comparable to peer vehicles, and would cost $137 million to modify with only a benefit of $49.15 million (based on the government's contemporary $200,000 value on human life and other associated costs) (p. 77). The costs and risks of some car crashes supposedly outweighed the potential damage of the poorly designed gas tanks. Although this could seem selfish on Ford's part, Velasquez reiterates that for utilitarianism, choices can't be for selfish reasons, and in this example, Ford alleged that not altering the design resulted in the greatest utility for society.

While benefits of this model are that it echoes principles of efficiency and attempts to reach the greatest good for all of society rather than isolated stakeholders, Velasquez reminds us that "benefits" of a particular action are not always equal to what is best for society, and that this model problematically relies on assumptions of quantitative methods and assumes that benefits and harms can be quantified and added and subtracted from each other to draw conclusions. As the Ford example illustrates, a financial compensation of $200,000 was assessed for the value of human life, and the costs and risks to change the poorly designed gas tank supposedly outweighed the risk of collateral damage. This model also diminishes arguments of rights and justice and insinuates that if costs are lower for society, then the rights and justice considerations for the societal outliers are not as important as the greater good (p. 86). In the case of the Ford example, its decision was based on the assumption that society would get the greatest "benefit" from an unchanged design, rather than taking the stance that it was more important in terms of rights and justice to allow customers with the potential to be injured to know about the problems.

Rights

The rights model is the second approach to moral and ethical decision making and deals with an individual's right or of freedom of choice. Velasquez defines a right as "an individual's entitlement to something" and can be used when referring to either how someone can act or how others should act toward that person (p. 93). If one has a right to free speech, then one has the right to speak what they want, and others are obligated to allow that person to speak freely. Thus, there are also both positive and negative rights, with positive rights being duties that others need to provide to the right holder and negative rights being duties to not interfere with others. Velasquez also details that there is a distinction between both legal and moral/human rights. Legal rights are rights granted by the legislative system, and moral/human rights are aligned with universal norms as to what people are entitled to as humans in general and not just due to specific legal protections.

Overall, a rights perspective will "override utilitarian standards" (p. 95), because it focuses on the individual rather than the collective society that the utilitarian perspective emphasizes. (Although here we can see that the "collective" for utilitarian and social justice widely diverge.) Although what is considered a right can change, it often takes more social unrest to alter the protective rights and sway the social norms about what is acceptable for the greater good. Further, the rights type of ethic uses a different lens from the utilitarian approach, where wrongdoing is a matter of some type of injury; for the rights concept, "a person's rights can be violated without the person being injured or hurt in any obvious way" (p. 93). For instance, Velasquez uses the example that one person can spy on another without the other's knowledge. No harm is necessarily done, but because the other values and has a right to their privacy, spying can be a violation of one's rights. This is particularly salient for surveillance and privacy, because as stated in chapter 1, surveillance can be a felt experience but one in which harm doesn't manifest itself with a "dead body."

Arguments of data protection and privacy fall into this rights-driven category, with data protection carrying the belief that "a balancing is required of the right to privacy, economic interests and individuals' well-being" (p. 370), and with privacy drawing on the belief that individuals should be able to live a life beyond an unwanted gaze.

As described, ethics of surveillance are often split into two categories that overlap with some of the conversations above: either (1) a rights-based approach or (2) a discursive approach. First, rights-based approaches position ethics as a matter of fundamental rights that should be granted by the state.

JUSTICE

The third ethical area Velasquez outlines is justice, which he links to fairness, but differentiates by aligning justice with more serious matters and aligning fairness with more fundamental beliefs about what is moral. The concept of justice covers the distribution of rights and costs in larger swaths of populations and moves beyond the individual like the rights category. However, principles of justice are based on individual rights and the "moral right to be treated as a free and equal person" (p. 106). Velasquez outlines that justice and fairness "are concerned with how one group's treatment compares to the way another group is treated" especially concerning aspects like benefits, burdens, laws, cooperation, competition, punishment, and compensation (p. 106).

Three conversations that justice is often divided into are distributive, retributive, and compensatory justice. Distributive justice looks at "distributing society's benefits and burdens fairly" (p. 107). Distributive justice often explains who gets goods or services, or what seems like essentials like "jobs, food, housing, medical care, income, and wealth." This area is concerned with debating the right way to allocate basic life needs because often either the demand is disproportionate to the supply, or there may be supply, but the quality of the supply may be less desirable. For instance, there may be work, but it might be unpleasant work, or there may be housing, but housing that is falling apart or substandard, or there might be health care, but coverage is limited or inadequate. Second, retributive justice makes sure penalties are deserved and fair when someone is blamed or punished for doing something wrong. This type of justice asks for the mental condition of the one committing the offense and whether the person either didn't know they were doing something wrong or that they couldn't freely decide to do it or not to do it. There also must be the determination that something done was wrong, and that the punishment severity meets the severity of the wrong. If one can determine that some type of incorrect conduct was committed, then the punishment should fit the wrongdoing. Third, compensatory justice deals with compensating someone who has suffered some type of loss. This can involve replacing property if something was destroyed or monetarily compensating someone for their damaged reputation (p. 118). Particularly pertinent for a discussion of justice ethics, and for TC, too, is the social justice turn and how a justice approach in ethics equates to a social justice approach in TC. It is noteworthy that Haas has called social justice a "new ethic" for TC (Walton et al., 2019, p. 175).

CARE

Finally, the ethics of care looks more specifically at individual relationships and takes a stance to care "for the concrete well being of those particular persons with whom we have valuable close relationships" (p. 121). This is a more personal approach to ethics and looks at stakeholders as individuals, rather than statistics or as vessels to facilitate the greater good. This stance sees the maintenance of relationships, both individual and communities, as integral to one's moral decisions. Velasquez gives the example that if one's parent and a surgeon (whom one wouldn't know personally) were drowning and one could only save one of the two, a utilitarian perspective may say you are morally obligated to save the surgeon because he or she has the potential for more utility. However, an ethic of care may say to save the parent to whom you are bound by sentiments often considered in ethics of care, which involves "compassion, concern, love, friendship, and kindness" (p. 121). These types of claims are based on fundamentals of the self and the belief that one can't exist without the relationships with others. Thus, "to whatever extent the self has value, to that same extent the relationships that are necessary for the self to exist and be what it is, must also have value" (p. 122). While there have been criticisms of unfairness in so far as care could lead to favoritism, especially in the workplace or that care can be tiring if one is always having to think of someone else, Velasquez brings out that this is just a reminder that all categories of ethics need to be considered with each other and not in isolation. A sense of balance among all groups can aid in making well-rounded, ethical decisions.

Context and Situation

Before looking at the application of ethics to Snowden's example, it is also important to understand the role of context and perspective in any given ethical situation. As Velasquez (2012) had noted with the first step of ethical analysis, to take ethical action, one needs to think about the context of the situation. Kupperman (1969) adds, "If you really want to appreciate this particular ethical problem, look at the circumstances." In an ethical evaluation, perspective is key to understanding ethics. Given that premise, in the context of my own analysis, I want to articulate that these four categories are assessed from two perspectives: a "pro" and an "anti" Snowden position, or two premises from divergent perspectives. Second, for that assessment, it is useful to understand the context that Snowden was working with when he helped facilitate the disclosure of classified information.

To those ends, I'll make brief comments about the surveillance programs Snowden was working with. As discussed briefly in chapter 1, key names of these surveillance projects released in the Snowden disclosures include PRISM, Tempora, Five Eyes, XKeyscore, and the Enterprise Knowledge System. These five programs are in no way representational of all the surveillance programs Snowden was working with, but they do provide a sample of surveillance work. With this just this small sample, all together, these systems and collaborations combine to create a larger surveillance schema designed to focus not on the individual but to gather large quantities of information through mass surveillance such as an accumulation of phone metadata. Systems like these are designed to take large swaths of information, identifiable or not, to stir around and sift together to see what patterns start to emerge. The information then informs additional actions based on presumptions of the future. The systems function to create widespread pools of information that can prove to be useful in instances of successful thwarting of potentially damaging actions, but they also prove to be complicated for those thinking of privacy, surveillance, and the nature of human rights—matters that begin to emerge by pairing the example of Snowden with categories of moral and ethical decision making. To those ends, these five programs can be described as follows.

PRISM

First, in brief detail, PRISM is a surveillance program designed to obtain communications from internet companies. According to *The Washington*

Post's access to classified documents, PRISM is the code name for the American program used by the NSA and the FBI, where the agencies can tap "directly into the central servers of nine leading U.S. Internet companies, extracting audio and video chats, photographs, e-mails, documents, and connection logs," so analysts can "track foreign targets" (Gellman & Poitras, 2013). At the time of Snowden's disclosures, this access provided so much information that the *Washington Post* also reported it was "the number one source of raw intelligence used for NSA analytic reports" (*The Washington Post*, 2013).

Tempora

Second, along the same lines, the program Tempora is the name of a program used by the UK's Government Communications Headquarters (GCHQ) to surveil internet activity. While Snowden was employed by the NSA (and not the UK government), his leaked documents revealed that the GCHQ shared information with the NSA through this program (MacAskill et al., 2013). One example of information intercepted and the partnership between the NSA and GCHQ was the surveillance of the "political leaders attending the 2009 London G20 summit" (Landau, 2013).

Five Eyes

Third, another collaboration revealed by the leaked documents is Five Eyes, a surveillance partnership between the US, Australia, Canada, New Zealand, and the United Kingdom that began in the 1940s (Ruby et al., 2017). This alliance was formed to transmit information between countries for collaborative purposes and formed what Snowden called a "pan-continental super-state in this context of sharing" (p. 353). The designations "AUS/CAN/NZ/UK/US EYES ONLY" or "FVEY" on Snowden's disclosed documents demonstrate the telltale signs that the documents are part of this consortium.

XKeyscore

As a fourth example, XKeyscore is another related surveillance structure and a computer system that assists in collecting internet information. As Fuchs and Trottier (2017) describe, XKeyscore is a surveillance system "that the NSA can use for reading e-mails, tracking Web browsing and users' browsing histories, monitoring social media activity, online searches, online chat, phone calls and online contact networks and following the screens of individual computers" (p. 413). It reportedly can also search both content data as well as metadata, which makes it a valuable tool for its strong searching capabilities (Greenwald, 2013b). As library and information scientists can attest, searching is one primary function of effective systems (McCulloch et al., 2005). Once information is gathered, as Lucas (2014) details, the program analyzes and categorizes swaths of metadata and "reviews these connections and assigns a resultant risk-analysis score." It allows the government to comb through correspondence on all types of platforms and technologies for goals like gaining "a decision advantage" (which is the published goal of the NSA) (National Security Agency, "Mission and Values," n.d.a).

Enterprise Knowledge System

Finally, fifth, according to Lucas (2014), the Enterprise Knowledge System is a series of related analytical and risk analysis databases that can group content, especially metadata (Lucas, 2014). *The New York Times* reported that a 2008 document revealed it had a $394 million multiyear budget and could "discover and correlate complex relationships and patterns across diverse data sources on a massive scale" to find targets (Risen & Poitras, 2013). Lucas (2014) comments, "Collectively, this system constitutes the means by which the enormous trove of so-called metadata is parsed, subdivided, and finally 'mined' or analyzed in accordance with the various legal regimes and presidential directives that govern the permissible use of this information." It is a system that assembles a variety of seemingly disparate pieces of information to identify patterns or inconsistencies and makes what may seem like unimportant information important. Lucas continues that through "data-chaining," "seemingly random, obscure, and even trivial information" is linked "into a variety of patterns" with the result much like "a kind of topographical mapping of patterns of movement and communication that filters out or excludes the metadata of most of us—unless that data can conceivably be gathered up into a pattern that is interesting . . . or suspicious."

Analysis

Now with that background in mind, one can move to an ethical analysis of the surveillance surrounding Snowden. Table 4.1 sketches the two

Table 4.1. Ethics, Snowden, and the NSA

	Utilitarian	
Description	An evaluation of the benefits and costs of choices on society.	
Perspective	**Pro-Snowden** Mass surveillance systems are costly for society because the drawbacks outweigh the benefits.	**Pro-NSA** Mass surveillance systems are beneficial for society because the advantages outweigh the costs.
	Rights	
Description	One is entitled to things either by law or just due to being human. People are obligated to allow others their rights.	
Perspective	**Pro-Snowden** Society is entitled to at least have transparent surveillance systems.	**Pro-NSA** Society is entitled to be safe, and nontransparent mass surveillance systems are acceptable because they help safety, security, and rights.
	Justice	
Description	Justice concerns the distribution of entitlements and fairness across society.	
Perspective	**Pro-Snowden** It isn't fair that the power of the mass surveillance systems is in the hands of an unaccountable few in a democratic society.	**Pro-NSA** Mass surveillance keeps society safer, which preserves democratic principles. Further, the programs are targeted, and through technology and algorithms, suspects are the targets, not average Americans.
	Care	
Description	Emphasis of personal connections and independence.	
Perspective	**Pro-Snowden** Someone with knowledge about mass surveillance should stand up for society and those they care about to expose the unaccountable powers.	**Pro-NSA** Mass surveillance for domestic and national security concerns is a way to ensure loved ones stay safe. Exposure of the systems is detrimental to the protection of loved ones.

Source: Author provided.

evaluated positions of each side of the argument paired with a different ethical perspective. Caveats are that though I set up the examples as two competing perspectives—pro- or anti-Snowden—it is important to reiterate that more than two stances exist, and binaries were constructed for illustrative purposes. Surveillance isn't an either/or position—it's gray, and it's fuzzy. One could theoretically support Snowden and the NSA simultaneously.

In addition, one's perspective can also shift at times, especially depending on current policies and current events. In fact, Snowden himself was a surveillance worker engaged in processes of mass surveillance (literally, supporting the surveillance industrial complex[3]), who slowly began to question his duties because he felt society wasn't getting a chance to debate the limits and extent to which the surveillance was being carried out. Snowden had enlisted in the U.S. Army under the justifications, "I wanted to be a liberator; I wanted to free the oppressed" (Snowden, 2019b), and he thought that being a veteran could eventually lead to government tech jobs to do so. He had wanted to work for the government doing surveillance work, but at least at some point, he decided too much secret surveillance was being conducted and he no longer wanted to participate. Snowden said that he eventually arrived at what he considered the act of whistleblowing, because he was torn between ideas of protection and concerns and the pro-privacy stance became a stronger urge.

UTILITARIAN ETHICS

Going into a little more depth for each of these ethical scenarios for mass surveillance, the first noted category of ethics was the *utilitarian* approach. A pro-Snowden stance could view the ethics involving the situation of surveillance systems as more detrimental than beneficial. This stance is best exemplified in Snowden's statements in *The Guardian* in 2013, when he discussed the circumstances around the start of his skepticism. In Geneva, Snowden was working with network security and was embedded with CIA employees. He reportedly grew more uncomfortable not only working with the computer systems but also in watching tactics used in the agency, especially when transforming contacts into spies. Snowden was quoted as saying, "Much of what I saw in Geneva really disillusioned me about how my government functions and what its impact is in the world . . . I realised that I was part of something that was doing far more harm than good" (Greenwald et al., 2013). Together with the previous statement that these surveillance systems put too much power in the hands of an "unaccountable few" (Gellman & Markon, 2013), Snowden's position was that many of the practices of government surveillance, including his involvement with them, were overwhelming detrimental society at large.

On the other hand, a pro-NSA approach could view the situation as the opposite, with societal benefits outweighing the drawbacks. Exemplifying this view are statements from governmental leadership. At the time of the disclosures, at a conference in Berlin, President Obama upheld that the programs were needed, and they served a beneficial function (Bruce, 2013). His comments referenced previous testimony by NSA director Gen. Keith B. Alexander, who cited that at least fifty threats had been thwarted and lives had been saved since 9/11, in part due to these mass surveillance systems (C-SPAN, 2013; Parkinson, 2013). Examples of attacks that were allegedly stopped included attacks against the New York Stock Exchange and the New York City Subway System. When supporting this stance, one would assume that society was better off with sustained surveillance to manage risks such as terrorism.

Overall, both positions illustrate why one would want to learn a literacy of ethics, as Cargile Cook (2002) argues for when recommending that students understand how to approach a situation in an ethical way. To put this in action, a communicator would benefit from approaching a situation considering three more important things Velasquez (2012) stated to remember about the utilitarian model. First, the choices must be good for society rather than just for the associated entity making the decisions (p. 79). A technical communicator could ask who is (not) benefiting from one's actions—the communicators themselves? A company? Society? Second, costs and benefits need to be considered not just in the immediate sense, but rather, all direct and indirect costs in the foreseeable future need to be identified. If a communicator is engaged in a practice of surveillance, is it okay in the short term? Might there be a long-term consequence? A pro-Snowden stance may see surveillance as causing long-term privacy, data-storage, and power-balance issues and question a prolonged state of exception legislation, such as the PATRIOT Act, which allows for short-term extensions of surveillance powers (as will be discussed in the next point). Finally, one must also compare other actions that could be carried out in the short and long term to determine that the present choice of action is the best in comparison to all other choices being presented. This third consideration would reiterate the need to think about other choices now and in the future.

Rights-Based Ethics

The second approach to ethics mentioned is rights based. Using a rights' approach to ethics requires a technical communicator to assume that one is entitled to something either by law or just due to being human. Not only does Snowden's illustrative example provide a heuristic for understanding a rights approach, but it also illustrates the distinctions and similarities between legal and moral/ethical considerations and shows how a range of rights considerations inform decision making.

Rights also come in different variants. Not only did Snowden and the journalists who published the information feel that (1) they had a right to disclose the information as whistleblowers, they also felt (2) that the government did not have a right to gather that information. Snowden asserted that from the Constitutional standpoint, the US had no right to engage in mass surveillance that would cast American citizens in the net. This is evidenced through his statement describing the purpose of his disclosures, "I gathered a lot of information about what I believed was evidence of criminal activity on the part of the United States government's unconstitutional programs" (Kegu, 2019). Further, in a CBS interview, Snowden posed the accusatory questions, "What harms the country? Is it a war built on lies? Or is it the revelation of those lies? Is it the construction of a system of mass surveillance that violates our rights? Or is it the revelation of that by the newspapers that we trust?" (Kegu, 2019). Correspondingly, one of the journalists involved in handling Snowden's documents, Glenn Greenwald, stated that not only did he have a Constitutional right to share the information, but he also had an obligation to tell the public. He reportedly stated, "As an American citizen, I have every right and even the obligation as a journalist to tell my fellow citizens and our readers what it is that the government is doing" (McCarthy, 2013).

On the other side of the spectrum, those supporting a government stance can also invoke arguments of legal and moral rights. While mass surveillance might not be an issue addressed specifically in the Constitution, the events of 9/11 acted as a catalyst for legislation that would help prevent future attacks and scaffolded some legal rights of mass surveillance. Prominently setting the stage for this surveillance and ethics clash was the familiarly named USA PATRIOT Act (long name Uniting and Strengthening America by Providing Appropriate Tools Required to Intercept and Obstruct Terrorism of 2001), which authorized "enhanced surveillance procedures" specifically focusing on "the interception of wire, oral, and electronic communications" (107th Congress, 2001). While there was "Congressional ambivalence" toward this bill (Farrier, 2007), many justified its means for the purposes of supposed increased security and risk mitigation. When signing the bill into law, President George W. Bush stated: "Surveillance of communications is another essential tool to pursue and stop terrorists. The existing law was written in the era of rotary telephones. This new law that I sign today will allow surveillance of all communications used by terrorists, including emails, the Internet, and cell phones. As of today, we'll be able to better meet the technological challenges posed by this proliferation of communications technology" (Bush, 2001). This legal Act was argued to be needed to protect the moral/human rights of American people, of which Bush declared, "Today we take an essential step in defeating terrorism, while protecting the constitutional rights of all Americans. With my signature, this law will give intelligence and law enforcement officials important new tools to fight a present danger."

Even two years after Snowden's revelations and thirteen years after 9/11, Congress reauthorized many portions of the Act through the USA FREEDOM Act (Uniting and Strengthening America by Fulfilling Rights and Ensuring Effective Discipline over Monitoring Act of 2015). Although the USA FREEDOM Act ended "the National Security Agency's disputed bulk phone-records collection program," it still maintained the existence of this information and put in its place "a system that will keep the records in the possession of phone companies" (Turner, 2015).

Not everyone agreed with the bill, but that wasn't necessarily because there was a disagreement about the surveillance practices. For some it did not adequately push boundaries—for instance, Senator Pat Toomey of Pennsylvania voted against the amended bill because it didn't go far enough. Using strong pathos and the threat of ISIS to support surveillance practices, he said, "They behead people. They crucify people. They burn people alive. They systematically sell young girls into slavery" with "their sights set on attacking the United States" (Turner, 2015). Overall, though, the USA PATRIOT Act and the USA FREEDOM Act are both examples of legislation that support the rights to government surveillance practices supposedly to ensure citizens can keep their legislated rights.

The spectrum of rights illustrates the range of possibilities for ethics in this area. Rights aren't just what one is entitled to do by law, but also what one feels obligated to do to, to sustain someone else's rights. This is a basic principle of the rights approach. First, Velasquez outlines "If I have a moral right to have someone do something for me, then that other person (or group of persons) has a moral duty to do it for me" (p. 95). This means that if one has a right to do something, other parties will either leave them alone to do it or will help them reach their goal. For instance, if one has a right to education, there must be a functional system in place to provide that education. Second, "moral rights provide individuals with autonomy and quality in the free pursuit of their interests" (p. 95). This means that individuals have a right to pursue their interests in so far as they don't need permission for their interests and their interests do not take from someone else. As Velasquez puts it, "To acknowledge a person's moral right, is to acknowledge that there is an area in which the person is not subject to my wishes and in which the person's interests are not subordinate to mine" (p. 95). Third, "moral rights provide a basis for justifying one's actions and for invoking the protection or aid of others" (p. 95). In other words, if one is granted the right to freedom of speech, then simultaneously one has the duty to make sure others have their own speech freedoms. This also means that individuals should have the ability to pursue their own interests, but also, it is morally correct to stop those who try to stop others from pursuing their rights. Several examples of various types of rights range from the legal rights granted by the U.S. Constitution or the Bill of Rights, to the workers' rights for equal pay and safe working conditions, to consumers' "right to know" such as the dilemma presented in the previous section about Ford's Pinto. There was support for the view that Ford should have been more upfront about the potential gas tank problems, so that consumers could have made the decision to purchase the car or not, "but they had no choice in the matter because they did not know the car carried this added risk" (p. 88).

In the case of Snowden's illustrative example, supporters of Snowden and the journalists he worked with could argue that Snowden acted ethically when disclosing the mass surveillance systems because the government did not have a constitutional right to collect information, and once aware of the systems, Snowden and the journalists had both a legal and moral right to let others know what was taking place. On the opposite spectrum, those arguing on the side of government could argue that the government had a legal and moral right to gather information and was granted that right through legislation granting constitutional exceptions such as the USA FREEDOM Act of 2015.

ETHICS OF JUSTICE

Third, in a justice approach, as discussed, the ethics of justice concern fairness and a distribution of entitlements across society. While rights deal with an individual's right, justice deals with the equal or fair distribution of those rights in a society. In the Snowden situation, one-way ethical principles of justice emerge through the continual reference to an entitlement of democracy, and with that, the assurance of government transparency. Snowden continues to reference that in a democracy, a just society needs to understand what the government is doing so that it can debate how they want to be governed. Democracy is thus framed as a matter of justice and entitlement, as Snowden keeps referencing that in democratic societies people are entitled to basic liberties such as transparency. Snowden reportedly stated, "I'm willing to sacrifice all of that [salary, career, and family] because I can't in good conscience allow the US government to destroy privacy, internet freedom and basic liberties for people around the world with this massive surveillance machine they're secretly building," and "What they're doing" poses "an existential threat to democracy" (Greenwald et al., 2013). Thus, without transparency (or too much government privacy), there is an unfair and unequal distribution of power, which means there is no longer a democracy. As stated before, Snowden had also justified his actions by saying the mass surveillance systems had put too much power in the hands of an "unaccountable few" (Gellman & Markon, 2013). Statements about the need for transparency are echoed by those who have served in positions of power within the NSA (thus showing that one can still engage in practices of mass surveillance but also feel the need for transparency). In a document posted on the NSA's website, in April 2000, Lt. Gen. Michael V. Hayden, USAF, and director of the NSA from 1999 to 2005, stated the following on record before the House Permanent Select Committee on Intelligence:

> In performing our mission, NSA constantly deals with information that must remain confidential so that we can continue to collect foreign intelligence information on various subjects that are of vital interest to the nation. Intelligence functions are of necessity conducted in secret, yet the principles of our democracy require an informed populace and public debate on national issues. The American people must be confident that

> the power they have entrusted to us is not being, and will not be, abused. (Hayden, 2000)

Hayden went further and argued that "opposing principles" of "secrecy on one hand, and open debate on the other" could "be reconciled successfully through rigorous oversight." By 2013, however, as told by Kegu (2019), Snowden seemingly decided that open debate wasn't happening and took it upon himself to take what he knew to the people to give the public opportunity for that oversight. Snowden reportedly concluded that, especially due to an eroding sense of public confidence in the media, "we're losing our position as a democracy and as a government that is controlled by the people—rather than people that are controlled by the government."

On the other side of the situation, mass surveillance is also framed as the instrument through which justice is maintained. The mass surveillance systems had initially been secured by the USA PATRIOT Act, sustained by the USA PATRIOT Improvement and Reauthorization Act of 2005, and many provisions were reauthorized or reorganized by the USA FREEDOM Act. The government defended the provisions of the Acts by assuring that the powers granted by the bills keep Americans safer, a safety everyone is entitled to. According to the George W. Bush's Whitehouse Archives, the Whitehouse's 2006 website when George W. Bush signed the USA PATRIOT Improvement and Reauthorization Act of 2005, provided the following support of the first two Acts:

> Since its enactment in October 2001, the Patriot Act has been vital to winning the War on Terror and protecting the American people. The legislation signed today allows intelligence and law enforcement officials to continue sharing information and using the same tools against terrorists already employed against drug dealers and other criminals. While safeguarding Americans' civil liberties, this legislation also strengthens the U.S. Department of Justice (DOJ) so that it can better detect and disrupt terrorist threats, and it also gives law enforcement new tools to combat threats. America still faces dangerous enemies, and no priority is more important to the President than protecting the American people without delay. (Whitehouse of George W. Bush, 2006)

The U.S. Department of Justice also published over 26 government officials' statements of support of the USA PATRIOT Act on their web page called

"Preserving Life and Liberty." For instance, Attorney General Alberto Gonzalez stated: "It is good in a democracy like ours that we discuss and analyze the wisdom of every law—particularly those that, if abused, would infringe your civil liberties. We have done that. Now, Congress must act to reauthorize the PATRIOT Act by sending the President a bill of which all Americans can be proud" (U.S. Department of Justice, n.d.). US Representative Bill Shuster also warned that America is under the threat of covert terrorists whose "end-goal is unclear but it does not include freedom or democracy." Thus, it is necessary for America to "defend its homeland by providing federal investigators with tools to track down terrorists within our borders," and "The Patriot Act provides these and is essential to winning the war on terror." Overall, though, as seen from leadership's support of these Acts, justice can be served not in absence of surveillance programs but through surveillance programs designed to maintain protections and uphold liberties for all Americans. All of America benefits through safety with sustained liberties. These positions compete with Snowden's comments that at least there must be a public debate as to whether the type of surveillance being authorized is desired by the collective whole. (It is of note that this is a very American-centric idea of justice, and although beyond the scope of this chapter, it would be worth looking at the rhetoric of justice for a global world from an intercultural communication stance.)

Before moving on though, it is worth mentioning that even though I related the justice perspective to calls for less surveillance due to justifications of societal democracy, when breaking ideologies down further, democracies obviously don't equate to across-the-board fairness and a just distributions of entitlements for all. What is considered "fair" and "just" are ideological constructs that differ with political philosophies within democratic societies. For instance, as Velasquez (2012) reminds us, theoretical schemes of distribution such as egalitarianism, capitalism, socialism, or libertarianism require particular views of people and goods in the contexts of terms like *freedom*, *governance*, or *competition*. An egalitarian position supports that all people are equal and should not be treated differently. Capitalists may feel that benefits should be proportional to contribution. Socialists may feel that all abilities and needs should be met. Libertarian positions may hold that no position is just or unjust—people should just be free to make their own choices (pp. 108–113). The purpose of this discussion was to at least provide an explanation of how mass surveillance was framed as a threat to democracy, or a right that needed to be distributed across all of society and not just allow a pocket of power in the hands of a few. However, it also raises questions of what justice is and for whom. This especially nods to TC's emphasis on social justice and the five faces of oppression, of which mass surveillance could be read especially as creating powerlessness or a form of cultural imperialism.

Ethics of Care

Finally, as discussed, ethics of care emphasize personal connections and focus on fostering not dependence but independence. From a position supportive of Snowden, someone with knowledge about mass surveillance would have concern for the well-being and independence of others and stand up for society to the unaccountable powers. Snowden echoes these positions when he reportedly states, "I really want the focus to be on these documents and the debate which I hope this will trigger among citizens around the globe about what kind of world we want to live in," because his "sole motive is to inform the public as to that which is done in their name and that which is done against them" (Greenwald et al., 2013). He continues, "I will be satisfied if the federation of secret law, unequal pardon and irresistible executive powers that rule the world that I love are revealed even for an instant."

On the other hand, there is also an argument for surveillance legalized initially through the USA PATRIOT Act, which pulls on rhetoric of ensuring others' happiness and assurances that loved ones will stay safe in the context of domestic and national security concerns. For instance, US Attorney General Alberto R. Gonzales spoke about the Act by saying Al-Qaeda and other terrorist organizations want to "demoralize our people" (U.S. Department of Justice, n.d.), which is an undesirable affect and positions the government's job as keeping its people encouraged. Further, justifying surveillance as a tool to fight terrorism, while at the same time assuring the public that the Acts have taken to account civil liberties, is taking a stance of care. *Liberty*, according to a dictionary definition, is the "quality or state of being free" regarding areas such as "the power to do as one pleases" and "the positive enjoyment of various social, political, or economic rights and privileges" and "the power of choice" ("Liberty," n.d.). In each of these areas, there is a running theme that a person with liberty gets to act as an individual who makes personal choices. So, when the government supports liberties and freedom, they are not only supporting rights and justice, but they are also supporting a type of care that means others will be able to act independently to make choices about their lives.

However, as shown, despite claims such as former Homeland Security Secretary Michael Chertoff's that "the PATRIOT Act gives us the ability to do that in a way that respects the Constitution, respects civil liberties, but gets the job done" (U.S. Department of Justice, n.d.), as shown, there is a larger debate about whether mass surveillance and powers granted by the USA PATRIOT Act actually uphold liberties or take them away.

There is a caveat about this, too, especially when dealing with Snowden's comments. In two ways Snowden and the government draw attention to the complexities about the ethics of care. When using care as an ethical heuristic, it is important to understand the motivation for the concern. Two comments by Snowden expressly illustrate this. First, as previously discussed, Snowden had stated that he was ready to sacrifice his family relationships because he thought the government was destroying "privacy, internet freedom and basic liberties for people around the world with this massive surveillance machine" (Greenwald et al., 2013). In this way, he showed an ethic of care for people around the world, but he also disregarded the more intimate familial relationships that were also important. At the same time Snowden stated he was concerned about the way others live in the world, he also admitted it was for more selfish reasons. After providing journalists with the sensitive information, *The Guardian* reported that Snowden stated, "I don't see myself as a hero, because what I'm doing is self-interested: I don't want to live in a world where there's no privacy and therefore no room for intellectual exploration and creativity" (Greenwald et al., 2013).

These comments emphasize a well-rounded examination of the motivations of care. While some of his statements seemed to show concern for others, some didn't. Velasquez (2012) highlights that when discussing concern, there are three ways to think about care: (1) caring about; (2) caring after; and (3) caring for someone. First, caring *about* someone involves caring either at a distance or about objects and ideas, rather than the more personal type of care involved in an ethic of care where one is invested in the well-being of the others. Second, caring *after* also denotes more of a disconnected relationship with something or someone else whose needs one tends to, but in a detached way, so that while one cares about the other's mortality, there is more objectivity and less of an investment in the personal relationship. A government welfare office could illustrate this position, when someone gets treated like a number but still receives life-saving care. Third, the type of care involved in an ethic of care, would be caring *for* someone, in the sense that one is interested in seeing through someone else's eyes and is invested and engrossed in one's "subjective reality" (p. 122). An example of this is parents caring for children, not to make them reliant, but rather to help them become independent decision-makers who can navigate the world to the best of their abilities. In these ways, then, some of Snowden's comments reflected the ethic of caring for others and having a genuine interest in their lives and subjectivity in the world, but his statements about self-interest reveal more investment in caring *about* others from a distanced and ideological space.

Moving Toward Action

Overall, these positions are particularly useful for seeing why oppression is only one way of perceiving surveillance. At the same time that someone like Snowden might frame surveillance as oppressive, others such as those who support mass surveillance might argue that the absence of surveillance can also lead to oppression. This is ultimately a question of rhetoric, where a pro-surveillance stance would invoke the surveillance-as-a-rhetoric-of-truth thread while anti–mass surveillance proponents might argue that transparency is truth, with both surveillance and transparency simultaneously making truth available through notions of visibility. This point is particularly useful for deliberation and creating dialogue between the two camps of people, and it helps us to think about both building consensus and dissensus, which can serve to make more visible "the crux of several fundamental disagreements between different factions" (Knievel, 2008). That these various positions can be made more visible themselves is a benefit of combining an ethical and just analysis to the situation.

Further, laying out these positions and using the framework of social justice and ethics, helps us to think about action. Bringing us back to the earlier discussion on socially just and ethical actions, while both ethics and social justice frames encourage action, they encourage different reactions. Further, if surveillance isn't seen as problematic, no action will be taken. This is a particularly salient point for TC and thinking about students as well, because recognizing spaces of surveillance and critiquing them are important. If we begin to see surveillance as oppressive, we can think about how to evaluate our own position, privilege, and power to start or join others in recognizing, revealing, rejecting, and replacing what we view as problematic. If we combine this with an ethical framework, we can think about how others might view the same situation as either not as harmful or not harmful at all. Breaking down positions into various pieces avoids oversimplification of ourselves and others and gives us an opportunity for critical reflection and space for change.

Further, this conversation also helps us realize that surveillance proliferates because not everyone views oppression as oppressive. While categorizing surveillance offers the possibility of addressing its oppressive nature, and thinking about surveillance can lead to forming a coalition of those who also realize problematic practices need to be rejected and replaced, even coalitional action won't necessarily stop surveillance. This is especially true because oppression can still benefit some people (often those in power), and, in particular, surveillance can even benefit the same people it oppresses. While surveillance by the state can cast a wide net disproportionally on certain communities or on those not suspected of a crime, many still argue that surveillance keeps communities safe (Office of Policy Development and Research, n.d.). Surveillance will continue to be accepted as long as someone is benefiting from it, and technical communicators as critical thinkers are poised to evaluate places of surveillance and to act when and where it is problematic.

Conclusion

To summarize this chapter's main objective, I argued that while social justice research can help us see that surveillance is oppressive, ethics as a heuristic can also help us see why someone would still support systems of surveillance. This was also important because although social justice and ethics aren't always positioned as allies, ultimately both frames encourage more ethical and just actions. Both ethics and social justice offer ways to react to surveillance, but an overview of both modes of action also helps explain why surveillance proliferates in the first place. Moving on to the next chapter, I will focus a bit more on how we can resist practices of surveillance once we identify uncomfortable or oppressive acts of surveillance for us or others.

Chapter Five

Resisting Surveillance through Tactical Communication and Social Justice

Building on the previous chapters, it's not only useful to identify the various dimensions of surveillance and how one can evaluate it, but it is also useful to see the agency involved in resisting one's role in scenarios of surveillance, especially if one identifies ethical concerns. This is particularly important to talk about when it comes to TC, because of a technical communicator's potential for action and potential to alter surveillance situations. As discussed in chapter 1, often popular surveillance narratives show that resistance involves paying more attention to privacy, and agency is illustrated from the target's point of view. If one is a subject of surveillance, then one can try to reduce visibility through privacy controls and conduct more cyber hygiene to keep one's data exposure minimized and protected (Cain et al., 2018). While useful, this position is also limited.

While targets of surveillance might be assumed to be those who resist with practices of privacy, as emphasized, the agents can also resist. This is an especially relevant point because this book has shown how technical communicators can be involved in carrying out surveillance, so thus the focus of this book has been on the agents (rather than the targets). Therefore, looking at how the agents resist is crucial for a more robust picture of resistance. Building from this conversation, this chapter explores two things: tactical communication and resistance through social justice. Tactics explore ways in which the individual can resist, and a conversation about social justice provides entry points into the ways that tactics can contribute to more systemic and collective forms of resisting harmful, oppressive, or just unwanted surveillance.

Tactics and Strategies

The first conversation is about resistance through tactics. To reveal systems of surveillance like Snowden did, in general, is tactical, and one could argue that Snowden's whistleblowing was a tactic to circumvent the strategies of the NSA. To explain this, one of the most useful ways to start the conversation is to explore the differences between tactics and strategies, distinctions that are laid out by de Certeau (1984) in *The Practice of Everyday Life*. In his work, de Certeau theorizes the difference between "tactics" and "strategies," and outlines that strategies are "the calculation (or manipulation) of power relationships that becomes possible as soon as a subject with will and power (a business, an army, a city, a scientific institution) can be isolated" (pp. 36–37). On the other hand, tactics are the ways that others react to those strategies and "make use of the cracks that particular conjunctions open in the surveillance of the proprietary powers" (p. 37). Simply put, for this conversation, strategies are the institutional ways of doing things, and tactics can be understood as the way those strategies are either carried out in practice or the ways that strategies are worked around.

The definitions of the terms highlight the centrality of the apparatus of the powerful institution in both strategies and tactics. De Certeau comments that a strategy "postulates a place that can be delimited by its own and serve as the base from which relations with an exteriority composed of targets or threats . . . can be managed" (p. 36). Further, he emphasizes that strategies can establish what is "proper" by mastering time through place and legitimizing knowledge with power. In other words, some type of organization, institution, and so on gets to say what is right and wrong by deciding what should be done in the space carved out as its own. Strategies capitalize on a place of power (or "the property of a proper") by using "systems of totalizing discourses" that are "capable of articulating an ensemble of physical places in which forces are distributed" (p. 38). Strategies are the actions of those in power.

When a powerful institution creates strategies through "logics, rules, or systems for organizing knowledge, genres, spaces, or bodies" (Colton et al., 2017, p. 61), the organization in power can also be said to encourage order and conformity, because it is advocating a particular way of being at the organization (de Certeau, 1984; Holladay, 2017). What the powerful entity does, and what it defines as important, becomes a standard or a norm to guide the thinking and actions of those under its power. For instance, employers set up systems for accomplishing work tasks as well as enact rules to govern employee conduct inside (and often outside of) work activities. A worker conducts business by engaging in certain procedures, communicates to others in particular formats, and does and does not share internal information in particular ways, such as Snowden's prohibitions not to share classified information.

For more flexibility to carry out one's duties (and to be less passive and reduce disempowerment), those subjected to strategies employ tactics to resist the strategies of power. Tactics become "the art of the weak" (de Certeau, 1984, p. 37), and as described by Colton et al. (2017), are "the various ways in which often individuals who are marginalized can appropriate strategies of control to suit their own ends" (p. 59). Tactics then, are the work-arounds to these official ways of doing things and are done by those with less power to enact strategies. Kimball (2006) comments, tactics aren't quite the sanctioned word on how things are supposed to be done, but rather, "Here's how I did it" (p. 74).

In those respects, tactics are also extra-institutional, as de Certeau (1984) notes, because what is created is in resistance and outside of the organization's strategies. A *tactic* (emphasis in original), he says, is "a calculated action" that exists as "a space of the other" (p. 37) and "must play on and with a terrain imposed on it and organized by a foreign power" (p. 37). That "foreign power" can be the strategies of the institution, and the tactic can be maneuvering around the official standards set out by an institution. For instance, employees may be prohibited from using their mobile phones during work hours, but may decide to take their phones into a closeted area or a restroom to avoid the detection of rule enforcers. In Snowden's case, he supposedly filled SD cards with materials that were not supposed to be shared and took them out of secure facilities by hiding the cards in Rubik's Cube, his sock, his cheek, and concealed in his pocket (Snowden, 2019a, p. 259).

What is inherent to tactics then is that they capitalize on space, time, and resources of the dominant strategy. De Certeau (1984) continues, saying that tactics can only take advantage of "opportunities" and "depends on them" as they emerge in the moment. Tactics "must vigilantly make use of the cracks that particular conjunctions open in the surveillance of the proprietary powers. It poaches in them. It creates surprises in them. It can be where it is least expected. It is a guileful ruse" (p. 37). Tactics then, also react in dominant spaces by taking pieces of strategies managed through surveillance and turning them into something different that can suit one's own ends. Yet, these pieces will never replace the whole; rather, they just react and work with or around it. De Certeau (1984) states, "A tactic insinuates itself into the other's place, fragmentally, without taking it over in its entirety, without being able to keep it at a distance" (p. xix). A tactic will be a reaction to the strategy, but because it's a reaction, it will always be connected and relegated to what it reacts to.

In the mid-2000s, Kimball (2006) popularized de Certeau's work of tactics in TC by describing the "tactical communication" (p. 67) in the form of enthusiast publications that car aficionados created to share their specialized knowledge with others outside of the institutions, a less extreme example (compared to whistleblowing) of tactics. In his work, Kimball (2017a) provided a general overview of the term when he described tactical communication as "technical communication conducted for reasons other than traditional institutional or strategic motivations" (p. 342). Instead of information being constrained by organizations, tactical communication exists in more distributed locations and reaches a wider audience. Kimball (2017b) states, "In effect, everyone who enjoys access to the Internet is now a potential technical communicator, sharing what they know about technology with the entire world" (p. 1).[1]

In addition to Kimball (2006), other scholars have taken up the topic with a wide range of issues that all stress technical work occurring outside the institutional setting. For instance, beyond Kimball's (2006) discussions of car devotees and their "enthusiast publications and their surrounding cultures" (p. 74); Holladay (2017) looks at discussion forums focused on technical documents related to psychiatric diagnosis; Colton et al. (2017) looks at Anonymous and the associated ethical implications of what can be considered (appropriate) resistance (p. 59); Seigel (2013) looks at more personal birth stories and pregnancy manuals; Sarat-St. Peter (2017) looks at terrorism and instruction manuals; Pflugfelder (2017) looks at Reddit and technical communication "in the wild" (p. 26); and Reardon et al. (2017) look at creators and players working with and beyond the confines of video games. While many of these focus on how the internet facilitates the sharing of technical communication outside of institutional contexts, at their heart, they all represent ways in which technical information is shared in unexpected, extra-institutional ways designed to share knowledge to others that would benefit from it.

It is also useful to note that there is a varied level of intensity for what can be considered *strategic* and what can be considered *tactical*. For instance, for strategies, while Ding (2009) refers to them in her example of SARS research as ways "to influence, guide, or manipulate human society" (p. 330), the reach of strategies doesn't have to be so grandiose. Colton et al. (2017) note that strategies, "can range from the benign and pragmatic (street signs, cross walks, grammar rules) to the manipulative (prohibitions on certain activities by certain people in certain spaces such as segregation)" (p. 61).

Similarly, tactics can range from the more ordinary offering of technical explanations of hobbies on online platforms like Reddit (Pflugfelder, 2017) to more radical attempts to share knowledge through guerrilla media to inform a surveilled and underinformed public (Ding, 2009). Kimball (2006) illustrates the variations of intensity in strategies and tactics when he talks about a traveler around a city. While one may be constrained by infrastructural strategies—"the layout of streets; the rules of the road; private properties; and economic, cultural, or governmental spaces"—the traveler may choose to deviate from that path because, "I'll go this way, it's prettier"; "I'll go that way so I can pick up some bread for dinner" (Kimball, 2006, p. 72). De Certeau (1984) reiterates the range when he states that many everyday practices like "talking, reading, moving about, shopping, cooking, etc." are characteristically tactical, but so are the "many 'ways of operating': victories of the 'weak' over the 'strong' (whether the strength be that of powerful people or the violence of things or of an imposed order, etc.), clever tricks, knowing how to get away with things, . . . joyful discoveries, poetic as well as warlike" (p. xix).

Thus, strategies don't have to be explicit directives of a political system—they are also found in the layout of a street (which admittedly is influenced by a political system) and cutting across a green lawn instead of using a sidewalk. Further, Colton et al. (2017) remind that a strategy is relative. The authors note, "A repressive 'order' means a specific, dominant political framework (e.g., capitalist, anti-individual autonomy) to de Certeau (1984)" but "to play devil's advocate, an order's degrees of repressiveness or oppressiveness can mean something entirely different depending on the perspective of the participants" (p. 62).

Tactics, Whistleblowing, and Snowden

Snowden's example of whistleblowing provides a useful illustration of how a technical communicator and Surveillance Worker like Snowden resisted institutional strategies by tactically providing journalists outside the United

States with proprietary information. In precedence, Ding (2009) has also made the connection between whistleblowing and tactics and international journalism in her work about the response to SARS in China. To begin with, whistleblowing can be considered tactical in at least three ways. First, whistleblowing is tactical because it is in response and resistance to institutional strategies. In a very basic definition, according to Markel (2009a), whistleblowing is "the act of publicizing illegal or unethical acts by an employer" (p. 126). Even more specific is Near and Miceli's (1996) popularly referenced position that "whistleblowing is the disclosure by organizational members (former or current) of illegal, immoral, or illegitimate practices under the control of their employers, to persons or organizations that may be able to effect action" (p. 508). Whistleblowing by definition becomes a rejection of the strategies of an employer.

Second, whistleblowing is tactical because it works within existing systems but finds the cracks that allow those with less power to make their own ends. As discussed, there is a powerful institution or discourse that represents the insider position and thereby creates an "other" that doesn't have that power; tactics of this other "must play on and with a terrain imposed on it and organized by the law of a foreign power" (de Certeau, 1984, p. 37) and make use of cracks that open up surveillance at the conjunctions of the powerful. Tactics are thus "the art of the weak" (p. 37) and allow marginalized to "appropriate strategies of control to suit their own ends" (Colton et al., 2017, p. 59). Whistleblowing then represents a way that an individual actor can step out of the dominant strategy to act in ways that might be contrary to the target's wishes (as a reminder, target in this case is the target of the whistleblowing, or the alleged wrongdoer). The NSA did not want Snowden sharing classified information.

Third, whistleblowing is also tactical because this approach is timely and represents a form of communication that can "concentrate the most [urgently needed] knowledge in the least time" (Ding, 2009, p. 342). Whistleblowing can produce those timely and adaptive characteristics by bypassing the traditional means of communication. Sarat-St. Peter (2017), drawing on de Certeau reiterates, "Tactics depend on a 'clever utilization of time': recognizing the 'precise instant' to intervene, responding to the 'rapidity of the movements' in an unfolding situation" (p. 86). For Snowden, these cracks were the affordance of international media. Instead of his superiors or even regulatory boards or local media, Snowden reached out to journalists. Ding (2009) also shows that the government's reporting of SARS in China was both tactical and whistleblowing and involved international media.

For the second assumption of the argument, based on these premises, Snowden could be considered a (at least self-identified) whistleblower, since, even if one disagrees with his reading of ethics and the Constitution, according to his statements as explored in the last chapter, Snowden extracted classified information and delivered it to journalists because he believed the US government was engaged in illegal activities of surveillance. Further, by utilizing de Graaf's (2010) seven key pieces of whistleblowing—actor, act, recipient, motive, subject, target, and outcome (p. 768)—Snowden (the actor) (debatably) blew the whistle (act) about the "mass surveillance of entire populations" (Snowden, 2019a, p. 1) (subject) against the US government and particularly the NSA (target) by providing classified materials to "journalists, who vetted and published them to a scandalized world" (p. 4) (the recipient), because he felt that helping to engineer "a system that would keep a permanent record of everyone's' life was a tragic mistake" (p. 3), and the "abuses I witnessed demanded action" (p. 8) due to ethical principles (p. 4), so much so that he felt morally obligated to break the orders as established by EO 13526 (p. 6) (motive). To him, the US had participated in the "most significant change in the history of American espionage—the change from target surveillance of individuals to the mass surveillance of entire populations" (p. 1). Snowden in particular picked journalists on the peripheries of the "fourth estate" (p. 246) who he thought "the national security state had already targeted" (p. 250). The outcomes of the event Snowden hoped for was "a return to the pursuit of the government's, and the IC's [intelligence community] own stated ideals," such as respecting a citizen's right to privacy on the internet (pp. 6–7) (outcome).

The outcomes Snowden did get was (1) both resistance and defenses of untargeted, mass surveillance from governments, corporations, and the public (Amnesty International, 2015); as well as (2) the US government's retaliation against him to announce charges under the Espionage Act (Siegel & Johnson, 2013); (3) charges contributing to Snowden's claim that he was forced into "exile" (Snowden, 2019a, p. 1); and (4) "time trying to protect the public from the person I used to be—a spy for the Central Intelligence Agency (CIA) and National Security Agency (NSA)."

Snowden shared information he was sworn not to share due to his TS/SCI clearance, or in his words, "strict compartmentalization under a legally codified veil of secrecy" (Snowden, 2019a, p. 238), and went to the foreign press because of what he interpreted as abuse of information he found unconstitutional. He went to an outsider through tactics because not only was his agency complicit in what he thought was troubling behavior, but that was the agency's regular form of behavior. It wasn't necessarily a rogue supervisor telling Snowden to lie about figures or participate in unsanctioned activities where Snowden could address an internal auditing team or even a third-party mediator. Snowden argues that what was being done at the NSA was systematically in direct opposition to constitutional affordances. Snowden's whistleblowing represented a more ideological impasse between what the Constitution granted and the methods the government was using to carry out its ends. (Important to note here, too, is that whether one believes Snowden was a whistleblower hinges on the degree to which one thinks what the NSA was doing was legal in constitutional terms.)

Complications to Consider

It is important to acknowledge, though, that even if one can argue that whistleblowing is tactical, there are some who do not think Snowden should be referred to as a whistleblower. As the US government's desire to charge Snowden with theft and Espionage Act violations (Esposito & Cole, 2013) points out, not everyone considers what Snowden did to be whistleblowing. This is a result of two things. First, there are disputes about what whistleblowing is to begin with. As de Graaf (2010) notes, the "difficulty in adopting general conclusions from the whistleblowing literature is that nearly every study's definition of whistleblowing is different" (p. 768). Second, Snowden's motives are also suspect, and motives are a contributor to whether something is whistleblowing. For instance, Near and Miceli (1996) note the distinction between informants and whistleblowers, saying that "informants reveal information, perhaps without thought to whether the recipient can use the information to stop the alleged wrongdoing," and whistleblowers "have some expectation that recipients are in a position to take some action with the potential of terminating the wrongdoing whether or not they finally do so" (p. 510). In his own work, Snowden (2019a) tried to differentiate between the more contemporary term *leaker* and *whistleblower* by stating, "Today, 'leaking' and 'whistleblowing' are often treated as interchangeable" (p. 236), with Snowden attempting to define a whistleblower (he argues, such as himself) as "a person who through hard experience has concluded that their life inside an institution has become incompatible with the principles developed in—and the loyalty to—the greater society outside it, to which that institution should be accountable"

(p. 238). Snowden says his motives stemmed from the government and intelligence community's acts of mass surveillance that in essence "hacked the Constitution" by thinking they were above the law (p. 233), but to some it isn't clear that his actions were ethically justifiable, which is again, returning to a circular discussion of what whistleblowing is and when it is justified. Overall, though, it is useful to have this conversation to show that whistleblowing can be a form of resistance to surveillance practices by capitalizing on tactical affordances. This particularly ties in Snowden's example to the conversation.

Beyond Whistleblowing for Tactical Communication

Whistleblowing is not the only form of tactical resistance available to a technical communicator, however; in fact, it might be even detrimental for creating spaces of resistance to think in such extremes. There are other more localized ways technical communicators can resist practices of surveillance beyond reaching out to international media, and one of those ways is through other forms of tactical (technical) communication. Technical communicators aware of their participation in surveillance can also share countersurveillance measures, privacy tactics, and other information to both agents and targets of surveillance in extra-institutional ways. Thinking back to the earlier conversation about tactical information in TC, there are various ways that one can also create some type of communication that works, but outside of institutional strategies. Examples include car enthusiast magazines, online forums, unofficial instruction manuals, and other online publications where one shares knowledge through unofficial means.

For technical communicators that identify themselves as agents of surveillance, one can choose to help others (including targets, other agents, or interested publics) see and resist surveillance scenarios by utilizing TC in various forms such as through Ding's more guerrilla methods,[2] delivered to those in a tighter social circle to more publicly in online forums or other extra-institutional channels as they arise. Context is what helps determine appropriate tactics of resistance, such as countersurveillance measures discussed by Marx (2003), including "discovery moves" to identify surveillance in the first place; "avoidance moves" such as suggesting one steer clear of certain surveillance spaces; or "blocking" or "masking" moves to maintain more privacy (pp. 374–380).

Examples of resistance can be online forums devoted to privacy or even on video platforms such as YouTube, where users can share their knowledge of surveillance systems to teach others about surveillance. For instance, a search of "technology to resist surveillance" comes up with a variety of responses seemingly from and for both those who do the watching and those who are watched. The YouTube channel *Rebbit* (2020) posted a computer-read "Ask Me Anything" conversation from the perspective of a former casino employee who responded to questions about casino surveillance regarding such things as using facial recognition and in spaces of high surveillance. In another example, Jerod MacDonald-Evoy (2017) posted his documentary explaining the pervasiveness and dangers of stingray technologies and what one can do about them. Both give insights into practices of surveillance that are delivered tactically and share specialized information in a more informal way. There is also a ready supply of others who hold more of the ethos of a content creator than an industry expert who shares tactics on the internet. A random search about digital privacy or surveillance yields pages of results by video creators who actively share tactics on the internet such as how to turn location settings off on a phone or how to minimize the exposure of one's data to Google (Young & Pridmore, 2021).

Snowden, too, has participated in the more localized, internet-based forms of tactical communication. Since his 2013 disclosures, Snowden has been more vocal online as a public figure speaking out against surveillance and detailing surveillance practices such as platforms like his Twitter page (Snowden, n.d.). Thus, Snowden is engaging in forms of tactical communication by taking his institutional knowledge of surveillance projects outside of the intended audience and providing useful tips for a more dispersed public.[3]

Everyday Tactics of Resisting Surveillance

Finally, it is also worth noting that one doesn't have to go quite so public with details of surveillance as an act of resistance. De Certeau (1984) especially spoke of resistance in the workplace when he described *la perruque* (p. 407), a tactic originally emerging from the workplace in which "the worker's own work [is] disguised as work for his employer" (p. 25).

De Certeau uses the example of an employee writing a love letter on company time or employees bringing a tool home to do work for themselves. The disguised work isn't stealing because no valuable materials are stolen, and it isn't absenteeism because a worker is still on the job; however, it does involve borrowing something or using something for one's own ends, be it company time to do something personal or borrowing a tool to use at home with no intention to take it forever, with emphasis that what is borrowed is used for something that is "not directed toward profit" (p. 25). When dealing with the element of time and the materiality of tools, this example is not only an instance of resistance, but also a resistance to surveillance, because at some point a worker wasn't engaged in the work they were supposed to be doing, or the worker left a tool or knowledge that was not supposed to be used in the way intended for one's own ends. The alternative use of time or tools becomes a tactical way to work around the employer's strategies, thus thwarting surveillant structures that might be in place such as time clock, a vault for materials, or computerized monitoring devices that monitor an employee's productivity. While the employee might not be able to break all rules lest they be fired, it is also possible to alter some of the prescribed ways of doing. Tactics thus become ways that systems of discipline and control are "vampirized" (p. 49).

For other forms of surveillance, resisting does not just mean one slacks off on the job or derelicts one's duties. One can also evaluate a surveillance scenario and choose the best course of action such as deciding not to engage in a practice or speaking to someone that might be the leading agent in the surveillance scenario and has abilities to make changes on surveillance practices. Taking smaller, concrete steps to reduce or remove harmful surveillance practices is also something to aim at. For instance, one can participate in the educational system but also note that it is inherently built on surveillance practices, some of which are problematic like the forced recording of students in their own homes for online, test-taking purposes.

It is very important to note, though, before continuing, that resistance in any form (e.g., whistleblowing, online forums, information sharing, resistance to everyday tasks) must be an ethical action, and it must be evaluated from a variety of levels. Further, this is no easy task as it has been repeated that surveillance is often a fuzzy, gray area that needs constant assessment. As Snowden also brings up, although TC may be able to facilitate resistance, it is not always desirable for everyone, and not everyone can agree on what is ethical. Colton et al. (2017) remind us, "[T]he line between ethical and unethical tactics can easily be blurred" (p.

62) in tactical communication (such as their analysis of the work from the group Anonymous), and there is a possibility that a tactic "intentionally wounds some with the intent of caring for others" (Colton et al., 2017, p. 71). Thus, it is still worth noting the power-upending potentials of tactical communication, but picking up from chapter 4, it is also important to consider the ethical considerations of not just what is being conducted (such as the surveillance actions taken by the government and the NSA), but it is also important to consider the ethical choices of divulging information that has been considered private.

Beyond being an avenue for communication resistance (such as the three ways that I have just described), tactics and tactical communication is also useful in one additional, reflective way. First, tactical communication is not only useful to perform resistance measures, at a metareflective level, the "tactical" represents a cultural[4] look at the values of not just an organization, but of a larger society in general. Kimball (2006) explains this for TC when he comments: "Examining technical documents and user communities such as these can reveal how people interact with technical documentation culturally—how deeply documentation can be integrated into the lives and fantasies of people in contemporary culture as they go beyond user-as-practitioner to user-as-producer and user-as-citizen. Observing such dynamics can provide some valuable insights about how technical documents convey and contribute to the forces of cultural power—whether centered around institutions, individuals, or dynamic communities" (p. 82–83). TC can thus benefit from seeing where there is resistance and where institutional actors can take note. Gilliom and Monahan (2012)[5] also explain this for surveillance when they comment: "[Resistance] is not simply a means by which hegemonic forms of discipline are tested and reinforced; nor is it merely a symbolic thorn in the side of totalitarian systems of oppression and control. Instead, resistance is ultimately generative and frequently self-affirming. Through resistance, people test boundaries, build sociality, and achieve dignity, both within and between institutional structures and dominantly cultural logics" (p. 407). For Gilliom and Monahan (2012), de Certeau's work is actually "optimistic," and resistance then is not only beneficial for those who are watched, but tactics are also beneficial for seeing larger patterns in what is or can become untenable. Through resistance, one can discover and/or critique how systems work and not just how systems are prescribed. A manager might be surprised to see that employees talk on phones in the restroom and could be motivated to alter break times in the employee's favor (or, more pessimistically, be more punitive).

Social Justice and Surveillance: Resistance and Replacement

Adding one more dimension to the conversation, resisting surveillance often needs more than tactics, however, and this leads back to the previous chapter's discussion offering steps for reacting to surveillance. Surveillance can be oppressive and is tied to hegemonies and power (Fuchs, 2011) that have the authority to watch and sort populations, often those more vulnerable to begin with. Dubrofsky and Magnet (2017) note that surveillance, especially for vulnerable bodies, "is always already bound up with gendered and sexualized ways of seeing" (p. 9).[6]

Oppression and social justice thus require that participants must be "collective" as well as "active" (Walton et al., 2019, pp. 50–56) because the type of change needed exceeds the power of one individual. Thus, while there is value in one person tactically sidestepping institutional strategies, there is strength in coalitional resistance. While tactics try to go around institutional strategies, social justice emphasizes challenging the whole system. That is not to say that tactics can't lead to structural change, but some tactics can look more individual, such as covering one's face to avoid being photographed, whereas coalition change might involve removing the camera or those who set up the camera in the first place. Tactics and social justice can also inform each other, too. In some ways, what Snowden did was tactical because he executed a work-around the strategies of his institution. In other ways though, he did call on the fundamentals of coalitions—he reached out to journalists to form a larger group with more ethos to give credibility and visibility to the documents he disclosed.

But of value here, too, is something not fully discussed in the last chapter and not fully addressed by Snowden. While Snowden made tactical and more collective moves that speak to three of Walton et al.'s (2019) 4Rs of collational action to recognize, reveal, and reject oppressive surveillance, he also did not robustly address the fourth R: that of replacement.

Replacement is a very important step in making change and can involve a substitution of "behavior, structures, or decisions" in either a personal or more comprehensive way (p. 142). Sometimes replacement can mean a personal behavioral change such as ceasing to say a word when one realizes it is offensive. For instance, the singer Lizzo rerecorded her song "Grrrls" when she realized a derogatory word could be considered ableist (Peiser, 2022). A more comprehensive change can require more resources to challenge power relationships, though, and possibly even restrict a whole organization through its policies and procedures (Walton et al., 2019, p. 142). This type of change is more dramatic and needs more coalitional action.

In some ways, Snowden did engage in steps of replacement. From a personal level, he quit his job and stopped the behavior that he deemed problematic. Instead, he went public with his identity and spoke out about his behavior and the disclosures. Further, on a more system-level, as noted, Snowden asked for more transparency, which could be interpreted as asking to replace the current system of transparency shrouded by a Top-Secret designation with something more transparent and open to debate. He argues that the government has a right to keep some information concealed, for example, "the identity of its undercover agents and the movements of its troops in the field" (p. 8), but not everything deserves secrecy, and the documents that journalists decided to make public[7] had to be published to let the people know what was happening. As quoted in chapter 1, Snowden stated, "The government and corporate sector preyed on our ignorance. But now we know. People are aware now. People are still powerless to stop it [sic] but we are trying. The revelations made the fight more even" (MacAskill & Hern, 2018).

While both stopping surveillance work and advocating for more transparency do represent potential replacements for current surveillance practices, Snowden could have gone farther with a more comprehensive plan for replacement. If he knows that there are so many stakeholders that either believe in the ethics of surveillance or are otherwise supportive of it for other reasons to include personal longevity,[8] a more thought-out plan for replacement could give more credibility for requests for changing practices. When organizations want to create change, they often turn to white papers or other documents that provide informative evaluations of possible change (Malone & Wright, 2018). Snowden also could have been more vocal about what to put in place of surveillance through some type of alternatives for implementation beyond transparency. This conversation would have spoken to Snowden's opponents, who would be supportive of surveillance but also don't know any other ways to operate.

It is difficult to say what Snowden "could" have done more of, however, without also acknowledging that a white-paper plan delivered by Snowden to replace larger-scale mass surveillance would be conceptually and philosophically challenging if not impossible and perhaps slightly laughable. As discussed, surveillance is rhetorical answer of truth, and replacing government-wide surveillance is not as simple as a list of recommendations. It also means rethinking of how we arrive at truth.

However, applying the idea of being prepared to offer an alternative to replace current systems in the smaller scale might be more realistic. For instance, if representatives for small retailer grew concerned with the amount of information the company was storing on users with their smart products, they could collectively develop an action plan on how to reduce this information gathering and storage, which could involve steps such as only keeping local data on smart readers or reducing the amount of time a smart reader sends information.

The Dutch government took similar steps when they rolled out their smart meter system and there was concern about the privacy-sensitive information the device would reveal. Cuijpers and Koops (2013) comment, "Real-time readings in intervals of minutes can reveal many details of home life and paint a disturbingly clear picture of people's behaviour and preferences" (p. 285). To minimize the data, the country decided there would be no central storage of meter readings, and there is a detailed code of conduct for how often information should be accessed (Van Aubel & Poll, 2019). While, again, this doesn't replace surveillance, it does demonstrate a more accessible example of a work-related surveillance issues that a technical communicator might face and the value of being able to offer solutions to replace surveillance practices.

Conclusion

Overall, in terms of both tactical communication and actions focusing on social justice, resistance to surveillance sometimes misses the target when it argues for being more private or being "cleaner" with one's data, as privacy paradigms often encourage (Bennet & Raab, 2007). As this book has argued, there are various levels and dimensions of surveillance, and everyday technical communicators may be involved in practices of everyday surveillance. Resistance then also focuses on the agency of the agent, and can and should also mean agents look inward to assess and reduce the spaces where surveillance may exist in the first place, especially if it is problematic. Resistance involves ethically evaluating scenarios, teaching others how to identify those scenarios, and considering where and how one can even resist, reject, and replace spaces of surveillance. This can be from a personal form of resistance, or it can be sharing with others how to resist. Resistance can be exceptional and visible, but it can also be quiet and "everyday resistance" (Gilliom, 2001; Gilliom & Monahan, 2012).

This also begs for future considerations, and another question arises: how do we prepare for a more surveillant future? One answer to that is in a TC classroom, and the next chapter will address how surveillance and TC are especially compatible through the concept of surveillance writing.

Chapter Six

Surveillance Writing

A Pedagogy

So far, this book has introduced surveillance, explored who does surveillance and how, provided strategies for evaluation and resistance, and now moves on in this chapter to explain ways to teach surveillance to others. I argue that one way to go about teaching others about surveillance is through the concept of *surveillance writing*. Surveillance writing explains how surveillance is written and carried out through communication practices and provides an entry point to assess where policies, practices, and assumptions about surveillance converge in a tangible, material form. If surveillance and its consequences can be hard to see, one tangible location of it is in the writing that surrounds it. Explaining how surveillance is carried out and why it is consequential to others can be a start of advocacy itself that identifies and reveals problematic spaces and helps equip others with frames, lenses, screens, and terms for seeing surveillantly in order to reject and replace potentially harmful or problematic surveillance practices.

Surveillance writing not only addresses foundational areas of surveillance in ways that are familiar to students, but it teaches literacies for surveillance. In this chapter, since surveillance is best understood contextually, I will also provide entry points into three additional illustrative examples beyond Snowden, where students can make their own ethical assessments. An eight-week course with sample cases for contextual analysis will be illustrated at the end of this chapter. A less condensed course of 16 weeks could unpack these eight weeks and spend two weeks on each element. The topics presented could also be thought of as modules rather than weekly plans, and room could be made to explore a TC-friendly deliverables such as a website or infographic to display the information instead of a more textual, written essay.

Surveillance Writing and Communication

In a basic definition, surveillance writing is writing that identifies, records, monitors, or stores information about identified groups of people, whether it is the breadth, limitations, or particularities of such actions. It is writing that defines, carves out, or explains methods or technologies of surveillance and either authorizes certain actions, already assumes those authorizations, or critiques practices of surveillance. It is writing that comments reflectively on how surveillance practices fit in with legal, ethical, and stakeholder expectations and the implications that these surveillance practices may have. It is a tangible product and a routine practice that can be assessed for its role in surveillance, and it is so common that it likely occurs daily, pulling from power, and drawing in both the willing and unwitting agents and targets.

INSPIRATIONS

The inspiration behind surveillance as a type of writing is drawn from the way scholars have positioned environmental writing as an area of focus which has its own genres and features. Johnson-Sheehan and Morgan (2009) note that *Time* magazine declared the 21st century the "Green Century" and add that many (to include technical communicators) have taken "an environmental turn" toward caring about climate change that has affected the global political and economic landscapes. This in turn has led to an increasing focus on writing that works to address the environment through conservation, or environmental discourse (e.g., Coppola & Karis, 2000), writing and ecocomposition (e.g., Killingsworth, 2005), ecological inquiry (e.g., MacMillan, 2012), food production (e.g., Richards et al., 2018), and spaces, places, and environmental justice (e.g., Goggin, 2013).

Similarly, surveillance has crept in as another dominant theme, and while the 21 century may have taken the "environmental turn," there has also been an embrace of surveillance and privacy that interrogates the way we are watched, how our information is managed and obtained, and how we resist. *Time* magazine even published work arguing how Snowden shows that a whole new "surveillance society"[1] is taking root (Von Drehle, 2013), and scholar John McGrath (2012) called the 2000s "the decade of universal surveillance," because there was now "the prospect of continuous, comprehensive surveillance" that was "accepted as a likelihood for most societies and a reality in many" (p. 83). In 2019, Greg Myre at National Public Radio called Snowden and his disclosures about mass surveillance one of the "stories of the decade." Taking inspiration from Johnson-Sheehan and Morgan (2009) then, this chapter focuses on surveillance *writing* as a way to address a larger phenomenon. I would also add that, as repeatedly stated and explored in chapter 4, surveillance writing is also closely tied to the social justice turn, which emphasizes critically evaluating and responding to potential and current spaces of oppression.

In conjunction to the definition of surveillance writing above, which classifies any type of writing and communication related to (or it could be argued as) surveillance, a surveillance writer can be someone who writes in any of the identified "*veillances" discussed in chapter 3, such as someone who works with software to gather intelligence information for the military, government, corporation, or for a hobby; someone who works with the data gathered by these systems; law enforcement officers who write reports; someone who works with medical surveillance by tracing the chain of infections such as with COVID-19; someone engaged in postmarket surveillance for medical devices and prescription drugs; a lawyer reviewing how a company keeps track of customers; a company keeping track of customers; an employee tasked with keeping track of customers; a reporter analyzing police use of body cameras or watching anyone in general; the HR department assessing employer and employee needs; the everyday worker who writes reports about quality assurance; or the manager who tracks the productivity of employees; and someone who writes codes to establish technologies that gather and track information.

Even machines can be agents involved in this writing in an actor-network kind of way.[2] If data collection is automated, machines "write" about surveillance through their collection of information. Machines can also adapt through deep learning practices. Even if data collection isn't automated, if technology is involved, it is at least an actant in a larger surveillance writing chain. Gallagher (2017) also suggests that algorithms are audiences, too. Overall, surveillance writing, at simultaneously its most basic and broadest, is writing that relates to surveillance, and a surveillance writer is one that carries out those activities.

It is useful to note that the idea of everyday surveillance is particularly important for surveillance writing, because while it might be easy to imagine the work activities of a Surveillance Worker to be someone sitting in a van and monitoring a fugitive with a lot of equipment, if contemporary surveillance practices are more of the everyday occurrence, then there also must be more commonplace practices of surveillance, too. Part of Staples' (2000) microtechniques of surveillance and the smaller disciplinary actions carried out through meticulous rituals of power can be writing practices that communicate information about or information taken through surveillance. As just mentioned, technical communicators can be susceptible targets when receiving a performance report, and they can be agents of surveillance if they are managers reviewing a performance appraisal. Even teachers of TC and grading are fundamental to surveillance, from both the standpoint of the grading process in general as monitoring as well as the storage and use of grading information for analytic purposes (Johnson, 2020). Surveillance writing affords the protections and/or outlines the constraints for both agents and targets in these scenarios.

Strategies

The work of Johnson-Sheehan and Morgan (2009) and their strategies of teaching conservation writing provide a particularly useful example for the inclusion of surveillance writing into TC pedagogy. It is of note to mention that the word *strategies* here is different from the way it is engaged by de Certeau earlier in the book. In this section, strategies are more general plans for accomplishing a goal rather than extensions of institutional power. To accomplish their goals, Johnson-Sheehan and Morgan outline several useful strategies, four of which I will discuss now. Their work mentioned: (1) learning about critical issues to look at how arguments are made and discourse is used (p. 22); (2) studying existing communicators already functioning in these spaces to identify possible roles; (3) practicing writing in the genres (p. 23); and (4) studying associated laws, politics, and the larger ecosystem surrounding the issues (p. 24). Similarly, these four steps can loosely be applied to concepts of surveillance and privacy.

Learn about Critical Issues

A first step for students then, can be to learn about critical issues. Students first need to be able to identify forms of surveillance writing in the first place. This can be achieved by teaching fundamentals of surveillance as well as professional and tactical materials. Surveillance studies literature is useful for setting the stage for scholarly discussions and viewpoints on surveillance. Various introductory texts like the *Routledge Handbook of*

Surveillance Studies (Ball et al., 2012), *Surveillance Studies: An Overview* (Lyon, 2007), *The Surveillance Studies Reader* (Hier & Greenberg, 2007), or *Surveillance Studies: A Reader* (Monahan & Murakami Wood, 2018) provide useful surveys in the field, including topics such as the difference between panopticism, Orwellian surveillance, or control societies. Scholarship also helps identify main themes of surveillance such as types of surveillance (ubiquitous, lateral, participatory, etc.), the importance of power and (in)visibility, social sorting, and social justice. Literature also helps identify sites of surveillance such as the government or corporations; actors of surveillance such as the state, commercial organizations, technology, oneself, or friends and family; and sites of activism and/or how the surveilled resist such as obfuscation, legal means, or artistic displays. A thorough discussion on technologies of surveillance is also important due to the connection between surveillance, technology, and technical communication. While those are just limited topics, they offer useful entry points into the conversation. TC and computers and writing literature discussed elsewhere in this book too, make useful primers for basic surveillance foundations.

One doesn't have to be an expert on surveillance studies to help guide students through informed discussions on surveillance (or privacy) though, because fundamental principles of watching and justice ground many conversations about surveillance. There are also various public-friendly research centers or media outlets that talk about surveillance and privacy. The American Civil Liberties Union (ACLU) (2020b), Privacy International (n.d.), and the Electronic Frontier Foundation (EFF) (n.d.) are three groups that take a critical stance toward the topics and provide content students can read. Educational hubs like Surveillance Studies Network (n.d.) or the Centre for Research into Information, Surveillance & Privacy (CRISP) (2020) also offer jumping-off points and actively participate in social media spaces. Other important sources of information are those that can sometimes take surveillance practices for granted, including military publications such as the *Air Force Magazine* (Air Force Association, 2020) and technology themed magazines such as *Popular Science* (2020) or *Popular Mechanics* (2020).

Studying Existing Communicators and Genres

Second, students can also study others to find their own agency in work involving surveillance and privacy by studying existing communicators. As stated, surveillance is a legal, ethical, and political matter, often technologically oriented and involving communication of technical information. To further understand their role and the role of other technical, scientific, and business matters in these activities, students can do two things. First, they can look at job announcements as shown above to get the gist of what this type of writing entails; then, they can analyze the employers, departments, and projects that these jobs relate to, to get a general sense of the ecosystems that these types of jobs occur within. As mentioned, some jobs involve working with intelligence, some can involve creating code, some can revolve around tracking populations, and sometimes these jobs require working with others who are focused primarily on issues of surveillance and privacy, such as a journalist at a privacy nonprofit might be—though other jobs may be more solitary, such as a supervisor in middle management in charge of making sure team production is up, with surveillance being an unspoken goal.

Second, students can become overall sensitized to surveillance and genres of surveillance writing to thus be informed employees who can recognize the hallmarks of this type of writing, which helps them either identify how their current work could be considered surveillance or privacy writing, or how they could provide space for themselves in future jobs that may not even exist yet. For instance, prior to the pandemic in 2020, some privacy jobs related to HIPPA and the collection of health data from employees didn't even exist yet (Chiavetta, 2021). This can lead to more critical analysis of one's job tasks, or it could even lead to more tactical actions such as Snowden's whistleblowing. In general, a more critically aware worker, supervisor, designer, engineer, and so on can make more ethical choices when presented with surveillance practices.

One useful way to see the possible genres is through a chart to help structure varying categories. As a caveat, the following categories are built on the assumption that borders are porous and relationships are spongy. As Bowker and Star (1999) classically note, a category "valorizes some point of view and silences another" (p. 5). Thus, it is important not to be constrained by categories and to keep in mind that all groups work with each other, especially the technological category, which weaves through all the other categories.

To help students connect with the genres as potential spaces they may find themselves in, table 6.1 outlines the different areas to consider, such as government sites, corporations, writings for the self, media work, technology, or health. It may be useful to break up government sites and

Table 6.1. Genres of Surveillance Writing

Site: State/Government/Military/Government Contractor	
Types of Writing	**Description of Agents and Targets**
• Intelligence, investigative, and police reports • Policy and legislation writing • Records—census, tax, military service, etc.	• State-led or -sponsored writing (which includes government contractors) that either gathers information to monitor and control those under the jurisdiction of the actor and/or discusses the limitations of that surveillance or the ways in which privacy is addressed.

Site: Corporations and Workplace	
Types of Writing	**Description of Agents and Targets**
• Employee monitoring such as performance reviews • Human resource documentation • Memos, emails, letters, newsletters about hard surveillance and privacy procedures or softer practices such as production reports* • Risk analysis • Policy reports • Grant applications • Product terms of service • Customer and social media monitoring • User experience and usability • Corporate websites	• Corporate/industry-led writing (not being conducted on the behest of the state), which either gathers information to monitor and control (to include consumer experience) those under the jurisdiction of the actor and/or discusses the limitations of that surveillance or the ways in which privacy is addressed.

continued on next page

---

*Staples (2000) describes soft surveillance as inconspicuous, such as a building layout in a store, and hard surveillance as more obtrusive and confrontational, such as a security gates at a store's exit.

Table 6.1. Continued.

Site: Self	
Types of Writing	**Description of Agents and Targets**
• Tactical communication • Social media • Personal webpage and managed content	• Individual writing to gather, share, or protect information (such as social media), as well as to share tactics to engage in or resist strategic use of surveillance information by more powerful entities, such as whistleblowing or other forms writing that could be considered sousveillance or countersurveillance.

Site: Media	
Types of Writing	**Description of Agents and Targets**
• Celebrity/entertainment journalism • Documentary/biographic writing • Investigative reports • Opinion writing • Political writing • Sports journalism • Science journalism • Trade journalism • Watchdog journalism	• Writing by media members that gathers information to monitor and control those under the jurisdiction of the actor and/or discusses the limitations of that surveillance or the ways in which privacy is addressed, to include share tactics to engage in or resist strategic use of surveillance information by more powerful entities.

Site: Technology	
Types of Writing	**Description of Agents and Targets**
• Code • Algorithms • Metadata • "Smart" data • Bot writing • Machine learning	• Writing by humans or machines for the purpose of information gathering to monitor and control those under the jurisdiction of the actor and/or discuss the limitations or resistance of that surveillance or the ways in which privacy is addressed.

Site: Health	
Types of Writing	**Description of Agents and Targets**
• National disease registries • Medical records • Occupational health reports • Postmarket surveillance reports	• A subset of other actor-led sites that specifically focus on writing about medical information gathered to monitor and control those under the jurisdiction of the actor and/or discuss the limitations of that surveillance or the ways in which privacy is addressed. It is especially useful to think of this section within the context of the 2020 COVID-19 pandemic, as well as consider medical surveillance in nonpandemic times.

Source: Author provided.

government contractors based on one's aims (for instance, Snowden's role as a government contractor also played a role in his access to classified information [Snowden, 2019a, p. 126]). Genres include reports, policies, websites, journalism stories, social media content, terms of service materials, performance reviews, human resource documents, codes, algorithms, bot writing, deep learning, and deep mediatization. Each area has a special type of focus that can be seen through assumptions of daily job tasks, and each contributes to thinking about the various ways surveillance is carried out through writing.

The information in table 6.1 does not comprise a comprehensive list, because it would be impossible to include everything that could be considered surveillance, and chapter 3 spent time addressing how to identify spaces of surveillance in a more organized fashion. It does offer a useful classification, though. More important than adhering to this list would be addressing basic heuristic questions such as: Does this carry out/limit/resist principles of surveillance (based on your personal definition)? If any of the answers could conceivably be yes, then it makes sense that what one was engaging in was surveillance writing.

Overall, each of these genres focus on the observation of groups and the use of the information obtained from those observations. While the classification of types of workers was previously discussed, it is also useful to see more specific examples. A quick internet search of a job website provides an introductory understanding of what surveillance writing does. Using the term *surveillance writing*, the first page of an online job forum ("Surveillance Writer Jobs," 2020) returns the following types of jobs:

- A technical writer for intelligence surveillance and reconnaissance (ISR) software and/or operations
- Writing for radar systems development
- Research intern with medial surveillance equipment
- Medical writing and public health surveillance
- Postmarket surveillance (e.g., medical devices and pharmaceutical drugs)
- Quality assurance inspections and persons to surveil contractor
- Retail jobs requiring the use of surveillance equipment like electronic article surveillance (EAS)

As shown from these jobs, the range and function of associate jobs differ but are connected to methods, authorizations, and limitations of surveillance. It is especially important to note that surveillance writing isn't limited to these jobs, and it is not wholly useful just to use the search term *surveillance writing*. As discussed in chapter 2 and throughout this book, surveillance workers can be anyone involved in surveillance processes, and if one is writing about the process of monitoring and control, then one is engaging in surveillance writing. One can begin to get a better sense of the world of surveillance writing, though.

PRACTICE WRITING DOCUMENTS IN VARIOUS GENRES

After identifying relevant areas where surveillance writing can occur, students can also practice writing through surveillance genres. A good start for students is the simple email written from the perspective of a boss wondering why an employee's production has been low. Students can even think about their own interactions within the university. Another simple exercise involves students looking at how they tailor their own social media posts thinking that certain audiences will be looking at their content (taking both the watcher's and watched person's perspectives). Having students translate more complex technological, science, or engineering-rated surveillance and privacy tactics and information to a more public-facing audience in popular media is also of value. More complex or targeted practices can analyze code writing and thinking about what a code or an algorithm sets out to accomplish, writing a report from the perspective of a statistician in a police department reviewing crime trends, creating a chart to relay the latest figures on who has a particular disease, or writing an assessment of a user experience researcher writing generally about how groups like a product or feel about the privacy controls of a technology (which is even more specific to surveillance themes). Having students themselves write also brings awareness to the ways they are watched and also the changes they make in themselves when they think someone is watching. Whether those changes result in small modifications of behaviors or create a more dramatic chilling effect is a point of analysis.

Familiarize Students with Relevant Legislation

A final strategy is to look at how privacy and surveillance are regulated. Agboka (2020) mentions that this strategy was employed in his sample course. Legislation about surveillance and privacy are plentiful. There is legislation that enables governments to gather foreign information, such as the Foreign Intelligence Surveillance Act (FISA), FISA Amendments Reauthorization Act of 2017 (Hatch, 2018), Executive Order 10450 requiring background investigations for security requirements for government employment (Executive Order 10450, 2016), or other various other information privacy or data protection laws like Europe's General Data Protection Regulation (GDPR), which aims to at least put into writing a uniform standard of data gathering (see the European Data Protection Supervisor, 2020).

It is also important to note a lack of legislation, too, and how the United States lacks a comprehensive data plan with some smaller jurisdictions creating their own standards. It is also important to understand the Freedom of Information Act or the Privacy Act of 1974 (see the U.S. Department of Justice, 2015), which also requires the government to provide some of the details about whether a request to the government is made and granted. Knowledge of redaction is also important, because even with requirements to share information, redaction can still be used be it legitimately (although "legitimacy" is of course a debated idea) due to certain exemptions as outlined by the FBI (Federal Bureau of Investigation, 2020), or as a tactic of obfuscation. Even thinking about FERPA and its provisions is useful. There are also academic centers that would be beneficial in these discussions such as the Center on Privacy & Technology at Georgetown (n.d.), the Brennan Center for Justice (2020), which partners with New York University Law, or the Berkeley Center for Law & Technology (2020) at the University of California, Berkeley.

Overall, though, it would be useful for students to explore the basic laws that both enable and curtail surveillance as well as describe privacy protections. This not only helps users know the limits they might possibly encounter in the future, but it also helps them understand their rights and provides a window into what already is being done in surveillance and privacy studies. It is also interesting to note again how legislation is often framed around protections of privacy, with the limitations of surveillance being more obscure.

A Note about Privacy Writing

Before offering a sample course, teaching surveillance through the frame of surveillance writing first requires a note as to why privacy is not included here with surveillance. Surveillance and privacy are closely intertwined and sometimes seemingly inseparable (e.g., one conducts surveillance at the expense of privacy). Due to these close ties, there is often a blurred boundary between the two terms, especially when it comes to writing. What some may term as writing about surveillance may be considered by others to be writing about privacy. For instance, on the same job site as described in the surveillance writing section, and on the same day, at the same time, searches for "privacy writing" jobs ("Privacy Writer Jobs," 2020) resulted in the following types of descriptions:

- Cryptography and Privacy Research Group intern
- UX Writer that supports privacy approachably for the audience
- Digital producer that understands rights of data privacy
- Professional writing tutor knowledgeable about the Family Educational Rights and Privacy Act (FERPA)

- Privacy writer (PIA) that reviews and analyzes privacy policies and procedures

Going back to the distinctions around privacy and surveillance, as shown by the results, there are two different job-type styles that focus on diverging categories but are still complementary and useful to see. The surveillance results revolve around conducting surveillance processes, and the privacy jobs focus on maintaining privacy. Both can overlap, but a separation is still important to maintain in the larger scheme of research. Throughout this book, as Snowden's example has shown, what some may consider the problematic execution of surveillance practices could simultaneously mean insufficient consideration of both legal and ethical privacy protections.

Another example is a standard privacy policy. In 2020, the social media app TikTok (2020) described its privacy policy in the US as follows: TikTok will collect users' account information when a user sets up an account, collect information "from third-party social network providers," gather technical and behavioral information during use of the platform, examine information in a user's messages, and take information from a user's phone book if a user grants access. Sample data collected by TikTok ranges from age, password, language, phone number, user-generated content, social contacts, and survey or sweepstakes information. TikTok also gathers automatically collected information like IP address, geolocation data, metadata, and cookies. Simultaneously, TikTok's privacy policies also address the "reasonable measures" it adheres to secure information and how it complies with legal policies about the protection of information.

Thus, in what is prefaced as a privacy policy, TikTok simultaneously describes their surveillance practices as well as defines what they consider privacy. This illustrates, as Bennett (2011) notes, that it is often difficult to understand where privacy discourse ends and surveillance discourse begins. Most of the time, the differentiation between the two often involves a nuanced look at a particular case. For TikTok, the privacy policy is thus a written document about surveillance and privacy despite the label implying it is just writing about privacy. For scholars, the framework for one's research (either surveillance or privacy) would likely impact the way to assess this policy as being an example of privacy writing or surveillance writing.

Overall, though it would be useful to look at "privacy writing" in this chapter, as well, but that is beyond the scope of this work. Although almost inseparable, these two terms are not the same. Surveillance and

146 | Working through Surveillance and Technical Communication

privacy are two different elements that work with and against each other. Privacy has its own breadth of academic literature with different foci. For future research then, it would be useful to also spend time thinking about a more robust description of privacy writing and possibly the differences between the two distinctions, but due to scope limitations of the book, that will not be an undertaking in this chapter.

SAMPLE EIGHT-WEEK COURSE

It is also useful to see these points in action, particularly through using Snowden as an illustrative example, as this book has consistently done. Thinking in terms of strategy and pedagogy, students could follow the flow of the book and use the illustrative example of Snowden to do four complementary things, which were adapted from Johnson-Sheehan and Morgan (2009): (1) learn about critical issues; (2) look at those already operating in these spaces; (3) practice writing documents in various genres; and (4) familiarize themselves with relevant legislation and look at the larger ecosystem and issues around of this type of writing. For the Snowden example, these steps require that one must understand how the NSA monitors groups; what surveillance is (and at least generally, privacy) and how Snowden argues for it; the historical dimensions of surveillance and privacy in the US and elsewhere; genres that the Snowden case covers—as well as be able to identify the legal, ethical, and political dimensions of the Snowden case.

Table 6.2 is an outline of this eight-week course on surveillance writing. Each week is broken down into four areas: driving question, activity, assignment, and readings that draws from principles of project-based learning that encourages a driving question, inquiry, authenticity, critique, reflection, revision, and a final product that could be presented beyond the classroom (Buck Institute for Education, n.d.).

In general, the above plan usefully addresses Johnson-Sheehan and Morgan's (2009) sampled pedagogical strategies by first identifying critical issues to look at, how arguments are crafted, and how discourse is used (p. 22). Each week revolves around examining both primary or secondary sources that identify critical areas for surveillance topics. Second, the format allows students to practice writing in the genres (p. 23), which is especially the case where students write a sample of two genres. Third, it helps students look at associated laws and politics surrounding issues (p. 24), and this connection is strongest where students analyze and compare legislation. Finally, fourth, the framework provides a space to learn

Table 6.2. Sample Course Schedule for Surveillance and Privacy Writing

	Week 1—Literacies of Surveillance
Driving questions:	What is surveillance, and what can the example of Edward Snowden teach technical communicators about surveillance?
Activity:	**Class Period 1:** Introduce the course and fundamentals of surveillance that are appropriate for grounding Snowden in studies of surveillance such as basic definitions and the diversity of what can be considered surveillance.

Class Period 2: Discuss how Snowden complicates the conversation on surveillance and reiterate that surveillance isn't just about the state and big brothers. There are a variety of agents from a variety of sites, some of it known, much of it ubiquitous. Practices of surveillance are carried out day-to-day by workers and can be everyday activities. |
| Assignment: | Students should summarize various theories and sites of surveillance and write a white paper with background and a reflection on the practices of surveillance, which they could be, or are already, engaged in or have been the subject of. The assignment allows for a less personal reflection to minimize feelings of being surveilled through this activity. |
| Sample Readings: | • Chapter 1 of this book, "Introduction to Surveillance and Technical Communication."
• Galič, M., Timan, T., & Koops, B. (2017). Bentham, Deleuze and Beyond: An overview of surveillance theories from the panopticon to participation. *Philosophy & Technology, 30*(1), 9–37. https://doi.org/10.1007/s13347-016-0219-1
• Haggerty, K. D. (2006). Tear down the walls: On demolishing the panopticon. In D. Lyon (Ed.), *Theorizing surveillance: The panopticon and beyond*. Willan Publishing.
• Lyon, D. (2002). Everyday surveillance: Personal data and social classifications. *Information Communication & Society, 5*(2), 242–257. https://doi.org/10.1080/13691180210130806
• Murakami Wood, D., & Wright, S. (2015). Before and after Snowden. *Surveillance & Society, 13*(2), 132–138. https://doi.org/10.24908/ss.v13i2.5710 |

continued on next page

Table 6.2. Continued.

	Week 2—Why Surveillance Matters and Assessing It
Driving questions:	Why does surveillance matter, and how can we assess it?
Activity:	**Class period 1:** Students are introduced various theories about why surveillance matters (both positive and negative) and critique the nothing-to-hide argument to assess its limitations. **Class period 2:** Students read about more classical business ethics and select an example of surveillance and explain how the various ethical approaches can identify a potential for help or harm.
Assignment:	Students practice with their visualization and create a chart in some type of word processing program such as Microsoft Word, which differentiates between various forms of ethics (and their potential for good or harm) in the context of their chosen example of surveillance.
Sample Readings:	• Chapter 4 of this book, "Evaluations and Responses: Social Justice, Ethics, and Surveillance." • Jones, N. N. (2016). The technical communicator as advocate. *Journal of Technical Writing and Communication*, 46(3), 342–361. https://doi.org/10.1177/0047281616639472 • Marx, G. (1998). Ethics for the new surveillance. *The Information Society*, 14(3), 171–185. https://doi.org/10.1080/019722498128809 • Richards, N. M. (2013). The dangers of surveillance. *Harvard Law Review*, 126(7), 1934–1965. • Solove, D. J. (2011). Why privacy matters even if you have 'nothing to hide.' *Chronicle of Higher Education*, 57(37), B11–B13. • Stoddart, E. (2012). "A surveillance of care." In K. Ball, K. D. Haggerty, and D. Lyon (Eds.), *Routledge Handbook of Surveillance Studies* (pp. 369–376). Routledge.
	Week 3—Surveillance and Privacy Writing
Driving questions:	Who is a surveillance writer, and what type of surveillance and privacy writing exists for technical communicators?

Activity:	**Class period 1:** Students review chapter 2 of this book, "Surveillance Workers and Technical Communicators," and analyze pieces of surveillance writing samples from a variety of sources to identify competing narratives about the role and limits of surveillance. Students contemplate what it would take to get various positions. For example, do the jobs require specialized knowledge? Security clearances? **Class period 2:** Students critique the differences between government surveillance such as Edward Snowden participated in, and other types of surveillance writing such as basic performance reviews or usability testing reports. This could focus on aspects of his work, tactics, involvement in business organizations, potential business writing, acts of resistance, and what he would have learned in a classroom, showing why it is useful to not only see his example but consider what he did/did not learn.
Assignment:	Students compose two genres of surveillance writing pieces and produce a reflection paragraph that assesses the functions and implications of both.
Sample Readings:	• Chapter 2 of this book, "Surveillance Workers and Technical Communicators." • Chapter 3 of this book, "Information, Technical Communication, and Surveillance." • Chesler, C. (2020, May 4). Coronavirus will turn your office into a surveillance state. *Wired.* https://www.wired.co.uk/article/coronavirus-work-office-surveillance • D'urso, S. (2006). Who's watching us at work? Toward a structural–perceptual model of electronic monitoring and surveillance in organizations. *Communication Theory, 16*(3), 281–303. https://doi.org/10.1111/j.1468-2885.2006.00271.x • *The New York Times.* (2020). Times topics: surveillance of citizens by government. https://www.nytimes.com/topic/subject/surveillance-of-citizens-by-government • Smith, G. J. D. (2012). Surveillance work(ers). In D. Lyon, K. D. Haggerty, & K. Ball (Eds.), *Routledge handbook of surveillance studies* (pp. 107–15). Routledge.

continued on next page

Table 6.2. Continued.

	Week 4—Resistance
Driving questions:	Why resist, and how can you resist surveillance through the frames of tactical communication and social justice?
Activity:	**Class period 1:** Students discuss why and how to resist through tactical communication. They explore the question, how can Snowden's example help us move beyond resistance through increased privacy? **Class period 2:** Students move toward a deeper discussion of ethics through the frame of social justice and discuss how to resist oppressive surveillance, calling back to the conversation about why surveillance matters.
Assignment:	Students experiment with ways of resistance and produce a deliverable that addresses resistance to a particular fictional or real-world surveillance example, such as a memo to an employer about a particular job duty, a YouTube video pointing out a harmful surveillance practice, or documentation of a plan for data justice.
Sample Readings:	• Chapter 5 of this book, "Resisting Surveillance through Tactical Communication and Social Justice." • Ball, K. (2005). Organization, surveillance and the body: Towards a politics of resistance. *Organization*, *12*(1), 89–108. https://doi.org/10.1177/ 1350508405048578 • Cinnamon, J. (2017). Social injustice in surveillance capitalism. *Surveillance & Society*, *15*(5), 609–625. https://doi.org/10.24908/ss.v15i5.6433 • Dencik, L., Hintz, A., and Cable, J. (2019). Towards data justice: Bridging anti-surveillance and social justice activism. In D. Bigo, E. Isin, & E. Ruppert (Eds.), *Data politics: Worlds, subjects, rights* (pp. 167–186). Routledge, • Jones, N. N. (2016). The technical communicator as advocate. *Journal of Technical Writing and Communication*, *46*(3), 342–361. https://doi.org/10.1177/0047281616639472 • Monahan, T., Phillips, D. J., Murakami Wood, D. (2010). Surveillance and empowerment. *Surveillance & Society*, *8*(2), 106–112. https://doi.org/10.24908/ss.v8i2.3480

	Week 5—Legislation
Driving questions:	What is relevant legislation for this type of work?
Activity:	**Class period 1:** In small groups, students research and summarize legislation from different industries and from different regions reflecting on scope, ethics, and affordances. Students compare results with their peers, debate who is protected by the measures and the effectiveness of the different industry and state approaches, and they present their findings to the class in a reflective discussion. **Class period 2:** Students look at the legislation that would be involved in the Snowden example and continue their debate.
Assignment:	Students create a poster that paraphrases and contrasts two different policies that can range from institutional or international—one institutional, the other international—and evaluate the surveillance scenarios they cover, identifying agent, act, site, target, motivation, information, surveillance paradigm, and consequences.
Sample Readings:	• European Data Protection Supervisor. (2020, June 24). *Homepage.* https://edps.europa.eu • Hatch, O. G. (2018, January 19). *Text—S.139— 115th Congress (2017-2018): FISA Amendments Reauthorization Act of 2017.* https://www.congress.gov/bill/115th-congress/senate-bill/139/text/eah (although Hatch, himself, is a controversial author, which could be a useful, secondary lesson in a topic like deconstruction). • Solove, D. J. (2004). Reconstructing electronic surveillance law. *George Washington Law Review, 72*(6), 1264–1305.
	Week 6—Surveillance Ecologies
Driving question:	What is the larger institutional and social system behind surveillance writing in technical communication?
Activity:	**Class period 1:** Students examine the example of Edward Snowden and think about the various actors and sites of surveillance present in an everyday situation. This can include associated technologies used in writing about and

continued on next page

Table 6.2. Continued.

	Week 6—Surveillance Ecologies (continued)
	carrying out surveillance. Students link back to ideas of social justice and connect the institutions to power and explain how these institutions set the tone for surveillance practices. **Class period 2:** Students review job advertisements and sketch the networks needed to carry out surveillance or ensure privacy is addressed. Students address technology and evaluate how it is handled by the user and how it bridges the user to surveillance goals.
Assignment:	Students create a chart to visualize different jobs, relationships, and connections needed to sustain surveillance and privacy practices.
Sample Readings:	• Donaldson, A. (2012). Surveillance and non-humans. In K. Ball, K. D. Haggerty, & D. Lyon (Eds.), *Routledge handbook of surveillance studies* (pp. 217–224). Routledge. • Hayes, B. (2012). The surveillance-industrial complex. In K. Ball, K. D. Haggerty, & D. Lyon (Eds.), *Routledge handbook of surveillance studies* (pp. 167–175). Routledge. • Turow, J., & Draper, N. (2012). Advertising's new surveillance ecosystem. In K. Ball, K.D. Haggerty, & D. Lyon (Eds.), *Routledge handbook of surveillance studies* (pp. 133–140). Routledge.
	Week 7—Replacement
Driving questions:	What would legal, ethical, politically aware, and just surveillance writing look like in the Snowden case and beyond?
Activity:	**Class period 1:** Students discuss what legal, ethical, just, and politically aware writing is in various contexts, drawing from all their research over the course. Students assess their thoughts on Snowden's actions individual and the role of larger networks of individuals in responding to surveillance practices.

	Class period 2: Students explore the case of George Floyd, Immigration and Customs Enforcement, and/or the practice of coding to evaluate the associated information surveillance.
Assignment:	Students create a rudimentary legal brief arguing for a surveillance and privacy policy measure for an industry of their choice, simulating what it would be like to manage surveillance and privacy. Students compare their work with others in similar and different spaces.
Sample Readings:	• Alperstein, N. (2021). *Performing media activism in the digital age.* Palgrave Macmillan. (Specifically, Exploring issues of social justice and data activism: The personal cost of network connections in the digital age, pp. 143–173.) • Bauman, Z., Bigo, D., Esteves, P., Guild, E., Jabri, V., Lyon, D., & Walker, R. (2014). After Snowden: Rethinking the impact of surveillance. *International Political Sociology, 8*(2), 121–144. https://doi.org/10.1111/ips.12048 • Beck, E. (2016a). A theory of persuasive computer algorithms for rhetorical code studies. *enculturation, 23.* http://enculturation.net/a-theory-of-persuasive-computer-algorithms • U.S. Immigration and Customs Enforcement. (2020, July 6). Broadcast message: COVID-19 and fall 2020. https://www.ice.gov/doclib/sevis/pdf/bcm2007-01.pdf • U.S. Immigration and Customs Enforcement. (2020, July 16). SEVP modifies temporary exemptions for nonimmigrant students taking online courses during fall 2020 semester. https://www.ice.gov/news/releases/sevp-modifies-temporary-exemptions-nonimmigrant-students-taking-online-courses-during
	Week 8—What Others Should Know
Driving question:	What would you teach others about surveillance?
Activity:	**Class period 1:** Students identify various approaches to an infographic and develop an outline of categories they would like to include if teaching others about surveillance, particularly in the field they expect to work in.

continued on next page

Table 6.2. Continued.

	Week 8—What Others Should Know (continued)
	Class period 2: Students create a best practice infographic based on their assessment of surveillance and privacy in the workspace of their chosen industry.
Assignment:	Students finalize a persuasive infographic that shares the various dimensions of surveillance and privacy based on their interpretation of the course. Infographics presented to an outside audience as students see fit.
Sample Readings:	• Jones, J. (2015). Information graphics and intuition. *Journal of Business and Technical Communication, 29*(3), 284–313. https://doi.org/10.1177/1050651915573943 • Scott, H., Fawkner, S., Oliver, C. W., & Murray, A. (2017). How to make an engaging infographic? *British Journal of Sports Medicine, 51*(16), 1183–1184. https://doi.org/10.1136/bjsports-2016-097023

Source: Author provided.

about how writers and scientists are already functioning in these spaces to identify possible roles for themselves and potential audiences for their work. This is practiced in the weeks that look at jobs associated with the Snowden events.

Additional Examples of Entry Points

It is also useful to say something about the examples of George Floyd, ICE, and coding listed in week 7, as additional places to think about resistance. As frequently stated, surveillance and privacy literacies need to be evaluated in context, and unpacking these examples with students would be a space for rich discussion. To explain each a bit more, in the summer of 2020, protests about police brutality were especially intense in the United States. In one particularly impactful example, after George Floyd died as a result of being unrelentingly pinned to the ground by a Minneapolis police officer, protests in Phoenix broke out demanding justice for Floyd. While various dimensions of this incident could be explored for the importance for surveillance and social justice, one particularly salient point for surveillance writing is that when the Phoenix Police Department arrested over 100 individuals in protests, they just provided a cut-and-paste, generic statement of the offenses supposedly committed by those arrested (Biscobing & Blasius, 2020). As deemed by the judge, though, a cut-and-paste statement does not sufficiently indicate probable cause for arrests. To justify the arrests, police would have needed specific details for each arrest, through which prosecutors could determine whether to press charges. Generic data with little detail was not sufficient.

This is important for two reasons. First, it shows a need for surveillance awareness in the writing process and a legal and ethical understanding of what types of personal information constitute as evidence. While officers may not be the first people to be considered technical communicators, their use of technologies and their manipulation of the cut-and-paste function can be considered a matter of communicating through technologies. Second, although protestors could view the decision to throw out the arrests as positive, this also means that the justice system is calling for more surveillance to get those details. Seeing how and why surveillance proliferates is an important function of critiquing surveillance and unpacking the power relationships inherent to surveillance scenarios. For the larger picture, this example shows how the intersections of surveillance and TC are important to interrogate, for now and in the future, especially when considering Walton et al.'s (2019) introductory call for more attention to social justice in TC, where the authors state, "to enact social justice as technical communicators, first, we must be able to trace daily practice to the oppressive structures it professionalizes, codifies, and normalizes." Even before considering the everyday power present in everyday surveillance, the everyday procedures of the police are inherently spaces of power, and the surveillance writing that takes place there is an especially important area to consider.[3]

A second example illustrating the connection between surveillance, social justice, and TC is a message from the U.S. Immigration and Enforcement (ICE) from July 2020. In response to the COVID-19 social distancing measures related to schools and requirements for students to attend in-person classes, the U.S. Immigration and Customs Enforcement (U.S. Immigration and Customs Enforcement "Broadcast," 2020) issued the message, "Broadcast Message: COVID-19 and Fall 2020," with the following directive for programs expecting to be fully online in fall 2020: "Students attending schools operating entirely online may not take a full online course load and remain in the United States. The U.S. Department of State will not issue visas to students enrolled in schools and/or programs that are fully online for the fall semester nor will U.S. Customs and Border Protection permit these students to enter the United States." While there were some differences for schools with hybrid formats, this put schools in unfortunate circumstances, because it was both unsafe and not possible or legal to schedule in-person classes and problematic due to the memo's directive that students not only *must* be attend in-person classes, but also if they were not, ICE noted: "Active students currently in the United States enrolled in such programs must depart the country or take other measures, such as transferring to a school with in-person instruction to remain in lawful status or potentially face immigration consequences including, but not limited to, the initiation of removal proceedings." Threatening removal of students for attending online programs and courses as a result of the pandemic and not representing either the school or the students' choices become not only issues of social justice in that government imposes a sense of powerlessness on the international students and on schools (although the schools obviously retain more power), but also it becomes a matter of TC and surveillance because this directive is a written communication of policy as well as a matter that provides general directions and also regards matters of technology, calling on schools not only to keep track of their students, but also to sort them into categories for removal. Attending school online is insinuated as something less legitimate than in-person classes, which also reflects tinges of surveillance in that requiring students to attend in-person ensures their visibility and otherness in the educational, institutional setting. Although this memo was amended,[4] it still brings to the forefront ways in which surveillance and TC intersect. Surveillance here involved the government watching the students, the schools watching the students, calls for teachers to watch students, and the interpretation of all the iterations of the memo in between.

Finally, a third example is seeing codes and algorithms as both technical communication and concerns of social justice. Beck (Beck, E., 2016) and Brock (2019) offer useful summaries for how codes and algorithms are forms of rhetoric and communication that carry implicit and explicit assumptions. Lindgren (2021) shows how codes can appear objective, but coding is really both "a technical process" as well as a "situated and relational writing activity," where coders write codes through a negotiation of both technique and human meaning-making (p. 117). Codes and algorithms, often associated with technical communication, then, are especially important for their built-in but often unchecked human biases. These concerns are illustrated by public spaces like using predictive algorithms in policing (Bennett Moses & Chan, 2018) to corporate spaces and the potential consequences of data mining by companies like Google such as filter bubbles and their contribution to the decline of democratic diversity (Dylko et al., 2017). Codes and algorithms are also particularly important spaces to interrogate for future reserach due to their perceived status as trade secrets (Carlson, 2017; Dylko et al., 2017).

Limitations and Conclusion

Overall, this chapter starts as both a conversation starter and a request for more scholarship to lay out what studies of surveillance writing can entail and hopefully help TC engage more with surveillance. Similar to what Killingsworth (2000) notes about conservation writing, where students learn "highly specialized and inscrutably difficult technical information generated by environmental scientists and engineers" in various genres and are required to "make the information clear and understandable to multiple audiences or stakeholders" (as cited in Johnson-Sheehan & Morgan, 2009, p. 26), surveillance writers must also navigate specialized and technical information generated by technology or surveillance workers to use for or against different segments of the population. This also requires genre knowledge to understand both the technical aspects of the writing as well as the human and social implications of these technologies. While the goal in conservation writing is to either outline how land or other pieces of the natural world are used to ensure the longevity of human existence, surveillance writing also works, not necessarily to conserve nature, but to either legitimize, question, or outline functionally how regimes of watching are carried out with possibly potentially detrimental or unequal treatment of others. The conclusion will wrap up the arguments in this book as well as reiterate why surveillance matters—not just because it is increasingly prevalent but also due to the consequences it has in the immediate sense and for the future.

Before concluding, though, it is important to reiterate three points about teaching using this chapter's frame of surveillance.

- Academics isn't always about focusing on every possible area that can be covered. Specializing is important, too, so choosing what areas to emphasize is useful. This is particularly true of an eight-week course, which would overwhelm students new to the topic if all things surveillance were covered. A course on surveillance writing could also incorporate other forms of surveillance communication that examine more physical techniques of carrying out surveillance, such as more physical watching, or discuss the more affective responses to surveillance; for this section, however, I focused on writing because it is one of the most likely ways a technical communicator would engage with practices of surveillance.

- Other disciplinary discourses not mentioned in much detail also offer more complex perspectives on surveillance. To name a few, in the current description and beyond, there are spaces for more work through discourses like rhetoric and composition such as the writing of code and rhetoric and the potential connections between persuasion and presentation. There are also more connections to TC through workers, technology, and design, as well as connections to the fields of information and computer science through work on classifications or studies of software. Science and technology studies can also offer discussions on ICTs, the philosophy of technology, or actors and networks. Sociology or geography offer useful discussions on the relationships with others and place, and media and communication can address platforms and digital discourses. Surveillance studies, too, have many more conceptual frameworks to engage in.

- Finally, like conservation writing that looks at what should "be preserved, defended, and made sustainable" (Johnson-Sheehan & Morgan, 2009, p. 11), surveillance writing can also address these matters. As more surveillance is systematized, there are more questions of what should be conducted, what practices should be defended, and what is "sustainable" surveillance (as in, how can we make surveillance "sustainable" in the social context?[5]).

Conclusion

Moving into the conclusion, it is useful not just to reflect on where this book has been, but also to reflect on where surveillance and surveillance scholarship in TC can go. This conclusion engages in a little futurism and will thus address where we're going through the overall driving questions asked in the introduction: How are technical communicators also surveillance workers, and why does this matter for TC and surveillance scholarship?

Where We've Come

To summarize what this book has covered thus far: chapter 1 explained surveillance, TC, and the intersection between the two; chapter 2 argued that there is a range to both surveillance and TC, and that both surveillance and the role of the technical communicator are becoming more important as more information is digitized; chapter 3 focused on information and paid special attention to how technical communicators can identify the work that they are doing as both more formal Surveillance work and less formal, everyday surveillance activities; chapter 4 discussed social justice and ethics; chapter 5 highlighted resistance; and finally, chapter 6 engaged in an argument about how to teach fundamentals of surveillance.

Overall, the book argued that what it means to be a "technical communicator" and surveillance worker is expanding in an increasingly more digitized, distributed, and technological environment. Similarly, surveillance as a concept has gone through its own transformations, first being traditionally theorized through restrictive readings of the state and physical locations of power but now to more distributed and technology-facilitated, rhizomatic areas of power that pop up in a variety of places and spaces, sometimes through highly imbalanced power structures (e.g., the state watching its citizens) to more balanced power structures (e.g., friends reading each other's social media posts). Further, whether one sees surveillance as a positive or negative practice is not universal, and only by seeing surveillantly through ethical and social justice literacies can one evaluate surveillance's contradictory spaces. In addition, approaching surveillance as a genre of writing helps us to extrapolate surveillance's multidimensional characteristics, as well as highlights spots where one may engage in surveillance and illustrates where humans in a surveillance machine can resist.

Building on what has been discussed thus far, I'd argue that we can create a surveillance action plan based on the complexities and dimensions of surveillance we've explored. That action plan looks like the following:

- **Step 1:** Recognize what surveillance is.
- **Step 2:** Think about surveillance as only one rhetoric to be employed.
- **Step 3:** Evaluate the systems where situations of surveillance are engaged.
- **Step 4:** Evaluate in a personal way how you are a participant in that space, and whether you are an active participant or if you are an outsider or the audience. Also, what is your positionality when evaluating that space?
- **Step 5:** Identify the other participants in the spaces and their surrounding contexts while they exist in that situation of surveillance.
- **Step 6:** Evaluate how surveillance could be benefiting or harming the participants through frames like ethics or social justice.
- **Step 7:** Identify which participants have the power to act and how they can act.
- **Step 8:** Identify a desirable replacement action if necessary, and determine how you will engage in a method for replacement (e.g., personal and tactical or systemic and coalitional).
- **Step 9:** Engage in the rejection and replacement of the surveillance.

- **Step 10:** Monitor the situation and its progress. Be aware of function creep, as not every instance of surveillance starts out that way.

While still advocating for flexibility, this action plan offers a more systematic way to go about recognizing and reacting to spaces of surveillance that build on informed critique and ethical and just actions.

Where Are We Headed?

Moving from the past and the present then, it is also useful to speculate on what surveillance futures could look like. To do this, I draw on futures research. According to Kreibich et al. (2011), "Futures studies are the scientific study of possible, desirable, and probable future developments and scope for design, as well as the conditions for these in the past and in the present" (p. 1). In a normative-prospective approach, futurists combine both facts and empirical research with imagination and creativity to speculate toward particular imaginaries of the future (pp. 16–17).

One such approach to that is thinking about the future in both apocalyptic and utopian ways. According to Goode and Godhe (2017), utopias can be thought of as desirable visions for the future, and dystopias can be feared visions of the future. Both imaginations "are vital" (p. 110), because utopias "can disrupt common sense assumptions about what is 'realistic' and challenge us to question whether and how we could rethink and reshape society" (p. 118). On the other hand, dystopias offer grim pictures of the future, full of fear, "against intensely bleak backdrops" (p. 119).

In a desirable or utopian vision of surveillance, surveillance becomes a rhetorical option rather than the default mode of operation, and if it is used, people get to choose their level of visibility, decide how their data is shared, and even own the data that is gathered. What happens with any data is subject to ethical and social justice evaluations. Surveillance isn't allowed for predictive analytics, and groups aren't categorized or lumped into "haves" or "have nots." People don't feel exploited, powerless, marginalized, culturally discriminated, threatened, or oppressed. They don't feel afraid to be creative or exercise rights, and they stand up and support each other's interests, which are diverse and can even be subversive.

In a more fearful and apocalyptic vision of the future, the opposite happens. There is no choice, but rather surveillance takes place where the powerful can watch whomever they want at all times with no limits to the data gathering or the use of that data. People feel powerless to reject or replace the systems. People are so afraid to take action, for fear of an undesirable categorization or violence, that they follow checklists and only read government propaganda. There are clear, overt divisions between categories of people, blatantly organized around traits like race, class, sex, and gender, and people grow fearful of associating with "undesirables" and become jealous of each other's categorizations. The marginalized are further relegated to the sides by increased costs due to their categories, and those who transcend borders are limited in movements and categorical suspicion.[1] The apocalyptic characterization may be more familiar than a utopian vision; as noted, we already see these types of dystopias in the book, *1984*, or the television series, *Black Mirror*.

Both viewpoints illustrate possible extreme consequences of surveillance. At best, surveillance is unessential, transparent, and isn't used to assess life choices. At worst, it instils fear, creates powerlessness, encourages conformity, and categorizes everyone while creating societal divisions. I would argue that rather than pivoting to either of those positions, with the help of technical communicators and others invested in a more equitable and just world, there might be a position in between utopia and apocalypse—sustainable surveillance, in the sustainable development sense, at the intersection of people, planet, and profits (Etezadzadeh, 2015), where there is a level of homeostasis such that what is economically viable and desirable meets what is socially acceptable and environmentally supportable (Brooks & Young, 2022). What sustainable surveillance would actually look like is an area itself worth further exploration, but investment in the societal impact of surveillance and speaking out to reject and offer replacement for unjust surveillance has the potential to balance at least the pursuit of profits, because surveillance economies have to react to social demands.

Final Words

Overall, surveillance, and the Snowden disclosures themselves, are a personal topic for me, and the idea that someone is watching has been a lifelong constancy. Growing up I was aware of parental power, and that someone would be watching, out of concern, both the good and bad that I did. As a teen, I was given extra attention and care by others who extolled the rhetoric of care and were concerned that my low-income, single-parent household was not enough to soon make me a productive adult

(Jackson, 2002). In early adulthood, I traded some systems of surveillance for another, working in the security clearance industry, where I not only watched over others, but was personally subjected to reinvestigations of my own background to make sure I didn't set off any "red flags."

After the Snowden documents were released in 2013, however, I turned toward an academic assessment of surveillance. How was the work that I did contributing to harm? I had always thought of my work as a public good, trying to make sure those who handled information were doing what they said they were, but at what point was I part of a larger, oppressive machine? I was also taken in by the notion of chilling effects. How does my behavior change when I think someone is watching? Why is my default to be a "good girl?" And what is a "good girl?" What gendered ideas would translate into being a "good woman?" And, if I wasn't always trying to please, would I be more creative? Would I take more risks? Or, am I actually more creative because I think others are watching me and will entertain what I have to say? I thought about these questions even as I wrote this book.

In the end, among many other things, surveillance can be positive, negative, insightful, incorrect, helpful, harmful, useful, irrelevant, complimentary, or contradictory. To answer the central question of why this matters for TC and surveillance scholarship, I would argue that as communicators working through and with technology and technical information, we need to think about what it means to do technical communication and to be a technical communicator, because information is an argument. A similar evaluation should go into thinking about surveillance work and the surveillance worker, because information that a technical communicator gathers through surveillance organizes lives in both beneficial and detrimental ways.

The ability to be a skillful and well-rounded communicator depends on the ability to see the agency and feel the efficacy we have (or don't have) when employing surveillance practices, particularly when we recognize it as harmful to ourselves and others. Technical communicators can also be surveillance workers, and surveillance is often initiated, justified, and sustained through data, information, and knowledge structured by TC in possibly routine, mundane, and everyday tasks. The more TC scholars talk about the connections, the more layers and depth that can be added to the surveillance conversation and the more prepared we will be to thoughtfully work with, or resist, the various systems of visibility, whether they are blatant forms of monitoring and control or more subtle ways of watching each other. Even if we don't think we're engaging in practices of surveillance, it is useful to consider how surveillance can disproportionally impact others, and we can also consider how being watched professionally and personally impacts the way we think and act. It also provides strategies for agency to work with and through surveillance in TC. This book thanks those scholars who have already engaged in this work and invites others to take up scholarship to help explore the deep and rich area of surveillance further.

Notes

Chapter One

1. Wise (2016) defines the surveillance imagination as "the collection of stories, images, ideas, practices, and feelings that are associated with surveillance at a particular point in time" (p. 4). The surveillant imaginary is not necessarily a reality and is instead more of what society imagines practices of surveillance to be. Lyon (2017) adds that what we imagine is also tied to what we accept as moral and ethical.

2. For instance, Bigo (2006) discusses that the "banopticon" entails crossing borders and the creation of categories of whose movements authorities do or don't restrict; Gill (1995) theorizes the "global panopticon" doesn't have to entail a centralized body of surveillance to be able to move coordinate resources and people, and De Angelis (2007) sees the "fractal panopticon" as "a mechanism of interrelated virtual inspection houses" that create a grouping of multiple panoptic associations (p. 196).

3. As Andrejevic (2012) explains it, ubiquitous surveillance is a prospect where "it becomes increasingly difficult to escape the proliferating technologies for data collection, storage, and sorting" (p. 92) and is related to ubiquitous computing where everyday objects possess the ability to compute but also remember information.

4. See McKee (2011) for more information about data brokers.

5. Ball (2019) notes that research on what Zuboff calls "surveillance capitalism" has been around for 20 years or longer in surveillance research, but Zuboff has furthered and popularized the term.

6. Helen Nissenbaum's (2009) work on "contextual integrity" argues that privacy depends on the context of information and what may be okay for some to know can feel like a breach of privacy for others.

7. There is also criticism of only measuring surveillance by its potential or actual harm. It is of note that Solove (2011) states that "[p]rivacy is not a

horror movie, most privacy problems don't result in dead bodies, and demanding evidence of palpable harms will be difficult in many cases." "Harms" will be explored more in chapter 4.

8. Peers watching peers on social media sites may know they are not getting the "whole truth."

9. TC is said to have taken a "social justice turn" (Walton et al., 2019, p. 7), where scholars highlight the power structures and distribution of goods and services, thinking not only about the audience as the receiver of information, but also as an audience that at times needs to be advocated with, particularly those that are marginalized and oppressed.

10. It is of note that privacy, too, can be a privilege, and who it is afforded to is also a matter of social justice and often comes from an economic advantage. This conversation is beyond the scope of this book, however, as it moves into privacy research.

11. Microtechniques of control correspond to Focault's microtechniques of discipline, which Lyon (2007) describes as techniques that "target and treat the body as something to be observed and tested" (p. 48).

Chapter Two

1. During an interview with the *South China Morning Post*, Snowden even reportedly took the job at Booz Allen Hamilton for the purpose of gaining access to this information (Lam, 2013).

2. For the sake of classification in this paper, I use the term *Technical Communicator* to also refer to jobs beyond a "technical writer," such as those listed by the STC, including writers and editors, indexers, information architects, globalization, and localization specialists, and so on. While some such as the U.S. Department of Labor (2018d) just collapse the term *technical writers* into *technical communicator* when they state, "Technical writers, also called *technical communicators*"), I use the term *Technical Communicator* to encompasses a larger breadth of occupations beyond the equivalent of a "technical writer."

3. Kimball (2006) encourages scholars to think in more robust terms when he notes, "[O]ne should expect to see less of the traditional genres represented in technical communication textbooks and more of the kind of technological narratives in which users describe their own implementation of tactics, their own version of the shared tactical story" (p. 74). Tactics will be explained further in chapter 5.

4. I use the term *surveillance scenario* to mean a situation that involves surveillance and is an assemblage of the agent, act, site, target, motivation, information, surveillance paradigm, and consequences. This will be discussed further in chapter 3.

Chapter Three

1. It is useful to nod to the larger conversation between the words *information* and *data*, as technical communicators and surveillance workers also work with data. While the words may be swapped for each other without much notice, *data* and *information* can mean different things. According to Holwell (2011), data is the "mass of facts" captured on something. These facts are taken to mean something, but it just hasn't been determined what yet or by whom. Morville (2005) adds that data and information change when the facts start to become interpreted and mean different things to different people. Morville clarifies that data is "a string of identified but unevaluated symbols," while information is the "evaluated, validated, or useful data" (p. 46). For the simplicity of this conversation, I will mostly be referring to information, but it is important to acknowledge the difference (especially for the ensuing discussion on surveillance's treatment of information), because surveillance workers and technical communicators are often engaged in transforming data into information either through sorting or the rhetorical process.

2. There is a rich history of rhetorical scholarship in TC, so rich that I acknowledge that a deeper dive into the generative and rich work of scholars is beyond the scope and limitations of this chapter and book.

3. There are several other paradigms in scholarship, but these were selected for exploration here due to their illustrative abilities toward the end of the chapter. For more paradigms, see Marx (2016, pp. 43–44).

4. According to Wise's (2017) reading of Deleuze and Guattari, an assemblage is "a becoming that brings elements together" and is used frequently in surveillance studies; it was discussed in chapter 1.

5. While the whistleblowing concept is not necessarily relevant to the categorizing, the process of identification of various elements in diverse scenarios is.

6. Marx (2012) might use the phrase "surveillance subject" here to identify the "person about whom information is sought or reported" (p. xxv).

7. As Deleuze (1992) argued, contemporary surveillance practices are designed to see how we compare to others as "dividuals" rather than see how we stand alone.

Chapter Four

1. Lockett (2013) argues, however, that surveillance "tends to function undemocratically during the pursuit of 'national security'" (p. 6).

2. For TC, Cargile Cook (2002) described ethical literacies as a student's "knowledge of professional ethical standards as well as their abilities to consider

all stakeholders involved in a writing situation" (p. 15). A student's awareness of justice would also take place along similar lines of knowing what social justice is and where it deviates.

3. For more information on the term *surveillance industrial complex*, see K. Ball & L. Snider (Eds.), (2013), *The surveillance-industrial complex: A political economy of surveillance*, Routledge.

Chapter Five

1. Two things are of note here: (1) in definitional terms, even the whistleblowing previously described can be considered not just tactical, but also tactical communication; and (2) tactical, extra-institutional communication is also a way to see technical communication in somewhat of a "tc" sense rather than a "TC" sense.

2. Ding (2009) refers to guerrilla media as "interpersonal communication technologies widely used by and easily accessible to the general public, including technology-assisted media such as mobile phone text messages, phone calls, Internet forums and chat rooms, as well as traditional word-of-mouth communication" (p. 330).

3. It is interesting to note that Snowden (2019a) himself is said to have learned many of his skills from what could be considered tactical, technical communication. Snowden says that he also would "page through tiny blurrily photocopied, stapled-together hacker zines with names like *2600* and *Phrack*, absorbing their antiauthoritarian politics" (p. 56).

4. Hoebel (1966) describes culture as "the integrated system of learned behavior patterns which are characteristic of the members of a society and which are not a result of biological inheritance" (p. 5).

5. Surveillance studies also use the work of de Certeau, but as Gilliom and Monahan (2012) note, "Surveillance Studies would probably be better served by turning to the work of Michel de Certeau, which—with few exceptions—has been neglected in the field" (p. 407).

6. Cultural identities, or "structural organizing mechanisms" can include, (but are not limited to) elements such as "race, class, gender, body ability, age, religion, time, nation, sex, ethnicity, space, and sexuality" (Ken, 2010, p. 41).

7. Snowden (2019a) reminds us that "the number of documents that I disclosed directly to the public is zero" (p. 8).

8. Again, Snowden (2019a) argues that many politicians support surveillance programs for their political livelihood and their alliances with their party or donors over their belief that these surveillance programs protect national security (p. 204).

Chapter Six

1. Surveillance studies scholar David Lyon (2001) helped popularize the phrase "surveillance society" in his publication "Surveillance society: Monitoring everyday life."

2. See the website for Bruno Latour (n.d.) for more in-depth readings and conversations about actor-network theory and science, technology, and society (STS) studies than can be explained here.

3. The link between writing, power, and the police is especially noted by Seawright (2017) in her work, *Genre of Power: Police Report Writers and Readers in the Justice System*.

4. ICE reissued a news release last updated on July 16, 2020, stating, "Due to COVID-19, SEVP instituted a temporary exemption regarding online courses for the spring and summer semesters. This policy permitted nonimmigrant students to take more online courses than normally permitted by federal regulation to maintain their nonimmigrant status during the COVID-19 emergency" (U.S. Immigration and Customs Enforcement, "SEVP," 2020).

5. One of the three pillars of sustainability is "social sustainability," which has been described by Brown et al. (1987) as "continued satisfaction of basic human needs," as well as other "cultural necessities such as security, freedom, education, employment, and recreation" (p. 716).

Conclusion

1. Lyon (2007) explains that categorical suspicion is "the sense that simply inhabiting a categorical niche is enough to attract suspicion" (p. 106).

References

107th Congress. (2001 Oct. 26). H.R.3162—Uniting and Strengthening America by Providing Appropriate Tools Required to Intercept and Obstruct Terrorism (USA PATRIOT ACT) Act of 2001. https://www.congress.gov/bill/107th-congress/house-bill/3162

Adey, P. (2012). Borders, identification and surveillance. In D. Lyon, K. D. Haggerty, & K. Ball (Eds.), *Routledge handbook of surveillance studies* (pp. 193–200). Routledge.

Agboka, G. (2020). Legally minded technical communicators: A case study of a legal writing course. *Journal of Business and Technical Communication*, 34(4), 393–414. https://doi.org/10.1177/1050651920932198

Agrawal, N. (2017, Jan. 17). There's more than the CIA and FBI: The 17 agencies that make up the U.S. intelligence community. *LA Times*. http://www.latimes.com/nation/la-na-17-intelligence-agencies-20170112-story.html

Air Force Association. (2020). Home. *Air Force Magazine*. https://www.airforcemag.com

Albrechtslund, A. (2008). View of online social networking as participatory surveillance. *First Monday*. https://firstmonday.org/article/view/2142/1949

Allen, I. J. (2019). Negotiating ubiquitous surveillance. *Screen bodies: An interdisciplinary journal of experience, perception, and display*, 4(2), 23–38. https://doi.org/10.3167/screen.2019.040203

Allen, L., & Voss, D. (1997). *Ethics in technical communication: Shades of gray*. Wiley.

Alperstein, N. (2021). *Performing media activism in the digital age*. Palgrave Macmillan.

American Civil Liberties Union (ACLU). (2020a, May 22). ACLU v. CBP—FOIA case for CBP and ICE records related to the use of cell-site simulator technology. ACLU. https://www.aclu.org/cases/aclu-v-cbp-foia-case-cbp-and-ice-records-related-use-cell-site-simulator-technology

American Civil Liberties Union. (2020b). Privacy and surveillance. ACLU. https://www.aclu.org/issues/national-security/privacy-and-surveillance

Amidon, S., & Blythe, S. (2008). Wrestling with Proteus: Tales of communication managers in a changing economy. *Journal of Business and Technical Communication, 22*(1), 5–37. https://doi.org/10.1177/1050651907307698

Amnesty International. (2015). *Two years after Snowden: Protecting human rights in an age of mass surveillance.* Amnesty International. https://www.amnesty.nl/content/uploads/2015/06/two_years_after_snowden_final_report_en_a4.pdf?x26848

Amoore, L., & De Goede, M. (2005). Governance, risk and dataveillance in the war on terror. *Crime, Law, and Social Change, 43*(2–3), 149–173. https://doi.org/10.1007/s10611-005-1717-8

Anderson, B. (2006). Writing power into online discussion. *Computers and Composition, 23*(1), 108–24. https://doi.org/10.1016/j.compcom.2005.12.007

Andersen, R. (2007). The Rhetoric of enterprise content management (ECM): Confronting the assumptions driving ECM adoption and transforming technical communication. *Technical Communication Quarterly, 17*(1), 61–87. https://doi.org/10.1080/10572250701588657

Andrejevic, M. (2002). The work of being watched: interactive media and the exploitation of self-disclosure. *Critical Studies in Media Communication, 19*(2), 230–248. https://doi.org/10.1080/07393180216561

Andrejevic, M. (2005). The work of watching one another: Lateral surveillance, risk, and governance. *Surveillance & Society, 2*(4), 479–497. https://doi.org/10.24908/ss.v2i4.3359

Andrejevic, M. (2012). Ubiquitous surveillance. In D. Lyon, K. D. Haggerty, & K. Ball (Eds.), *Routledge handbook of surveillance studies* (pp. 91–98). Routledge.

Andrejevic, M., & Gates, K. (2014). Big data surveillance: Introduction. *Surveillance & Society, 12*(2), 185–196. https://doi.org/10.24908/ss.v12i2.5242

Angwin, J., Larson, J., Savage, C., Risen, J., Moltke, H., & Poitras, L. (2015a, Aug. 15). NSA spying relies on AT&T's 'extreme willingness to help.' ProPublica. https://www.propublica.org/article/nsa-spying-relies-on-atts-extreme-willingness-to-help

Angwin, J. Savage, C., Larson, J., Moltke, H., Poitras, L., & Risen, J. (2015b, Aug. 15). AT&T helped U.S. spy on internet on a vast scale. *The New York Times.* https://www.nytimes.com/2015/08/16/us/politics/att-helped-nsa-spy-on-an-array-of-internet-traffic.html

Aristotle. (1990). Rhetoric: Book I. In P. Bizzell & B. Herzberg (Eds.), *The rhetorical tradition: Readings from classical times to the present* (pp. 151–60). Bedford Books of St. Martin's Press.

Association for Career & Technical Education. (2005). Computer support specialists and systems administrators: Job description. *Techniques, 80*(7), 40.

Ball, K. (2005). Organization, surveillance and the body: Towards a politics of resistance. *Organization, 12*(1), 89–108. https://doi.org/10.1177/1350508405048578

Ball, K. (2019). Review of Zuboff's The age of surveillance capitalism. *Surveillance & Society, 17*(1/2). https://doi.org/10.24908/ss.v17i1/2.13126

Ball, K., Haggerty, K., & Lyon, D. (Eds.). (2012). *Routledge handbook of surveillance studies*. Routledge.

Ball, K., & Snider, L. (Eds.). (2013). *The surveillance-industrial complex: A political economy of surveillance*. Routledge.

Banville, M., & Sugg, J. (2021). "Dataveillance" in the classroom: Advocating for transparency and accountability in college classrooms. *ACM Conferences*. https://doi.org/10.1145/3472714.3473617

Barajas, J. (2016, Feb. 20). *How the Nazi's defense of 'just following orders' plays out in the mind*. PBS. https://www.pbs.org/newshour/science/how-the-nazis-defense-of-just-following-orders-plays-out-in-the-mind

Barton, B., & Barton, M. (2016). Modes of power in technical and professional visuals. *Journal of Business and Technical Communication, 7*(1), 138–162. https://doi.org/10.1177/1050651993007001007

Batova, T., & Andersen, R. (2017). A systematic literature review of changes in roles/skills in component content management environments and implications for education. *Technical Communication Quarterly, 26*(2), 173–200. https://doi.org/10.1080/10572252.2017.1287958

Bauman, Z., Bigo, D., Esteves, P., Guild, E., Jabri, V., Lyon, D., & Walker, R. (2014). After Snowden: Rethinking the impact of surveillance. *International Political Sociology, 8*(2), 121–144. https://doi.org/10.1111/ips.12048

Bauman, Z., & Lyon, D. (2013). *Liquid surveillance*. Polity.

BBC News. (2014, Mar. 18). Snowden in new US surveillance claim. https://www.bbc.com/news/world-us-canada-26640228

Beck, E. (2013). Reflecting upon the past, sitting with the present, and charting our future: Gail Hawisher and Cynthia Selfe discussing the community of Computers & Composition. *Computers and Composition, 30*(4), 349–357. https://doi.org/10.1016/j.compcom.2013.10.007

Beck, E. (2016). A theory of persuasive computer algorithms for rhetorical code studies. *enculturation, 23*. http://enculturation.net/a-theory-of-persuasive-computer-algorithms

Beck, E., & Hutchinson Campos, L. (Eds.). (2020). *Privacy matters: Conversations about surveillance within and beyond the classroom*. University Press of Colorado. https://doi.org/10.7330/9781646420315

Beck, E. N. (2015). The invisible digital identity: Assemblages in digital networks. *Computers and Composition, 35*, 125–140. https://doi.org/10.1016/j.compcom.2015.01.005

Beck, E. N. (2016). Writing educator responsibilities for discussing the history and practice of surveillance & privacy in writing classrooms. *Kairos: A Journal of Rhetoric, Technology, and Pedagogy, 20*(2). http://kairos.technorhetoric.net/20.2/topoi/beck-et-al/beck.html

Beck, E. N., Grohowski, M. G., & Blair, K. L. (2015). Subverting virtual hierarchies: "A cyberfeminist critique of course management spaces." In J. Purdy and D. N. DeVoss (Eds.), *Making space: Writing instruction, infrastructure, and multiliteracies*. University of Michigan Press. http://dx.doi.org/10.3998/mpub.7820727

Bell, L. (1994). The language cop. *Technical Communication, 41*(2), 275. https://www.jstor.org/stable/43090324

Bennett, C., & Raab, C. (2007). The privacy paradigm. In S. P. Hier & J. Greenberg (Eds.), *The surveillance studies reader*, 337–353. McGraw Hill Open University Press.

Bennett, C. J. (2011). In defense of privacy: The concept and the regime. *Surveillance & Society, 8*(4), 485–496. https://doi.org/10.24908/ss.v8i4.4184

Bennett, K. C., & Hannah, M. A. (2022). Transforming the rights-based encounter: Disability rights, disability justice, and the ethics of access. *Journal of Business and Technical Communication, 36*(3), 326–354. https://doi.org/10.1177/10506519221087960

Bennett Moses, L., & Chan, J. (2018). Algorithmic prediction in policing: Assumptions, evaluation, and accountability. *Policing and Society, 28*(7), 806–822. https://doi.org/10.1080/10439463.2016.1253695

Berkeley Center for Law & Technology. (2020). *Welcome students: What BCLT offers*. Berkeley Law. https://www.law.berkeley.edu/research/bclt

Bigo, D. (2006). Security, exception, ban, and surveillance. In D. Lyon (Ed.), *Theorizing surveillance: The panopticon and beyond* (pp. 46–68). Routledge.

Biscobing, D., & Blasius, M. (2020, June 2). *Phoenix police arrest dozens with copy-and-paste evidence*. ABC 15 Arizona. https://www.abc15.com/news/local-news/investigations/phoenix-police-arrests-dozens-with-copy-and-paste-evidence

Bitzer, L. F. (1968). The rhetorical situation. *Philosophy and rhetoric, 1*, 1–14.

Bivens, K., Cole, K., & Heilig, L. (2019). The activist syllabus as technical communication and the technical communicator as curator of public intellectualism. *Technical Communication Quarterly, 29*(1), 70–89. https://doi.org/10.1080/10572252.2019.1635211

Bizzell, P., & Herzberg, B. (Eds.). (1990). *The rhetorical tradition: Readings from classical times to the present*. Boston: Bedford Books of St. Martin's Press.

Blumer, H. (1954). 'What is wrong with social theory?' *American Sociological Review, 18*, 3–10. https://www.jstor.org/stable/2088165

Boedy, M. (2017). From deliberation to responsibility: Ethics, invention, and Bonhoeffer in technical communication. *Technical Communication Quarterly, 26*(2), 116–126. https://doi.org/10.1080/10572252.2017.1287309

Boehme, G. (2018). *Edward Snowden: Heroic whistleblower or traitorous spy?* Cavendish Square Publishing.

Boerman, S. C., Kruikemeier, S., & Zuiderveen Borgesius, F. J. (2021). Exploring motivations for online privacy protection behavior: Insights from panel data. *Communication Research, 48*(7), 953–977. https://doi.org/10.1177/0093650218800915

Bonk, R. (1998). Writing technical documents for the global pharmaceutical industry. *Technical Communication Quarterly, 7*(3), 319–327. https://doi.org/10.1080/10572259809364634

Bose, N. (2020). Amazon's surveillance can boost output and possibly limit unions—study. Reuters. https://www.reuters.com/article/us-amazon-com-workers-surveillance/amazons-surveillance-can-boost-output-and-possibly-limits-unions-study-idUSKBN25R2L1

Bowker, G. C., & Star, S. L. (1999). *Sorting things out: Classification and its consequences*. MIT Press.

Brennan Center for Justice. (2020). *Home*. Brennan Center for Justice. https://www.brennancenter.org

Brock, K. (2019). *Rhetorical code studies*. https://doi.org/10.3998/mpub.10019291

Brooks, C. F., & Young, S. M. (2022). The importance of absence: Information, surveillance, and UN's 2030 Agenda for Sustainable Development. Presented at the biennial Surveillance & Society conference of the Surveillance Studies Network, June 1–3, Rotterdam, The Netherlands.

Brown, B. J., Hanson, M. E., Liverman, D. M., & Merideth, R. W. (1987). Global sustainability: Toward definition. *Environmental Management: An International Journal for Decision Makers, Scientists and Environmental Auditors, 11*(6), 713–719. https://doi-org.eur.idm.oclc.org/10.1007/BF01867238

Bruce, M. (2013, June 19). NSA dragnet 'saved lives,' Obama says. ABC News. https://abcnews.go.com/Politics/obama-nsa-dragnet-saved-lives/story?id=19434444

Bruno Latour. (n.d.). *Home*. http://www.bruno-latour.fr/index-2.html

Brunon-Ernst, A. (2012). *Beyond Foucault new perspectives on Bentham's Panopticon*. Ashgate.

Buck Institute for Education. (n.d.). *Gold standard PBL: Essential project design elements*. https://www.pblworks.org/what-is-pbl/gold-standard-project-design

Burke, K. (1941). *Philosophy of literary form: Studies in symbolic action*. Internet Archive. Louisiana State University Press. https://archive.org/details/philosophyoflite00inburk/page/110/mode/2up

Burke, K. (1945). *A grammar of motives*. University of California Press.

Burke, K. (1990). Language as symbolic action. In P. Bizzel and B. Herzberg (Eds.), *The rhetorical tradition: Readings from classical times to the present* (pp. 1034–1041). Bedford Books of St. Martin's Press. (Original work published 1966).

Burley, H. (1998). Does the medium make the magic? The effects of cooperative learning and conferencing software. *Computers and Composition, 15*(1), 83–95. https://doi.org/10.1016/S8755-4615(98)90026-3

Bush, D. (1994). Being a friendly "language cop." *Technical Communication, 41*(4), 764–765.

Bush, G. W. (2001, Oct. 26). *Remarks on signing the USA PATRIOT ACT of 2001.* govinfo. https://www.govinfo.gov/content/pkg/WCPD-2001-10-29/html/WCPD-2001-10-29-Pg1550.htm

Cain, A., Edwards, M., & Still, J. (2018). An exploratory study of cyber hygiene behaviors and knowledge. *Journal of Information Security and Applications, 42*, 36–45. https://doi.org/10.1016/j.jisa.2018.08.002

Campbell, B. (2013, Nov. 1). *Ten talking points the NSA uses to justify its spying.* The World. http://www.pri.org/stories/2013-11-01/ten-talking-points-nsa-uses-justify-its-spying

Careers: Current openings. (2018, May 9). RealPage. https://rn21.ultipro.com/REA1005/JobBoard/JobDetails.aspx?__ID=*AA909AE4E336D0D0&__jbsrc=EFE6B008-FB32-45CB-972F-D366F5FEE2FD

Cargile Cook, K. (2002). Layered literacies: A theoretical frame for technical communication pedagogy, *Technical Communication Quarterly, (11)*1, 5–29. https://doi.org/10.1207/s15427625tcq1101_1

Carliner, S. (2012). The three approaches to professionalization in technical communication. *Technical Communication, 59*(1), 49–65.

Carlson, A. M. (2017). The need for transparency in the age of predictive sentencing algorithms. *Iowa Law Review, 103*(1), 303–330. https://ilr.law.uiowa.edu/print/volume-103-issue-1/the-need-for-transparency-in-the-age-of-predictive-sentencing-algorithms/

Carradini, S. (2020). A comparison of research topics associated with technical communication, business communication, and professional communication, 1963–2017. *IEEE Transactions on Professional Communication, 63*(2), 118–138. https://doi.org/10.1109/TPC.2020.2988757

Cassidy, J. (2013, Aug. 20). Snowden's legacy: A public debate about online privacy. *The New Yorker.* http://www.newyorker.com/news/john-cassidy/snowdens-legacy-a-public-debate-about-online-privacy

Cathcart, M. (1997). Strategic planning in a government organization: The experience of the technical information division at NRaD. *Technical Communication, 44*(4), 406–411.

Cayford, M., van Gulijk, C., & van Gelder, P. H. A. J. M. (2015). All swept up: An initial classification of NSA surveillance technology. In T. Nowakowski, M. Mlynczak, A. Jodejko-Pietruczuk, & S. Werbinska-Wojciechowska (Eds.), *Safety and reliability: Methodology and applications* (pp. 643–650). Routledge.

Center on Privacy & Technology at Georgetown. (n.d.). *Center on Privacy and Technology.* https://www.law.georgetown.edu/privacy-technology-center

Central Intelligence Agency (CIA). (2018, Apr. 30). *Careers & internships.* Central Intelligence Agency. https://www.cia.gov/careers/opportunities/cia-jobs

Centre for Research into Information, Surveillance & Privacy. (2020). *Home.* CRISP. http://www.crisp-surveillance.com

Cheek, R. (2020). Zombie ent(r)ailments in risk communication: A rhetorical analysis of the CDC's zombie apocalypse preparedness campaign. *Jour-

nal of Technical Writing and Communication, 50(4), 401–422. https://doi.org/10.1177/0047281619892630

Chesler, C. (2020, May 4). Coronavirus will turn your office into a surveillance state. *Wired.* https://www.wired.co.uk/article/coronavirus-work-office-surveillance

Chiavetta, R. (2021, Mar. 5). *Expect the privacy job market to stay strong, even after the pandemic subsides.* International Association of Privacy Professionals. https://iapp.org/news/a/expect-the-privacy-job-market-to-stay-strong-even-after-the-pandemic-subsides

Cinnamon, J. (2017). Social injustice in surveillance capitalism. *Surveillance & Society, 15*(5), 609–625. https://doi.org/10.24908/ss.v15i5.6433

Clark, C., Murfett, U., Rogers, P., & Ang, S. (2012). Is empathy effective for customer service? Evidence from call center interactions. *Journal of Business and Technical Communication, 27*(2), 123–153. https://doi.org/10.1177/1050651912468887

Clarke, R. (2016, July 24). *Roger Clarke's 'privacy introduction and definitions.'* Roger Clarke's Web-Site. http://www.rogerclarke.com/DV/Intro.html

Coll, S. (2014) Power, knowledge, and the subjects of privacy: understanding privacy as the ally of surveillance. *Information, Communication & Society, 17*(10), 1250–1263, http://doi.org/10.1080/1369118X.2014.918636

Colton, J. S. (2016). Revisiting digital sampling rhetorics with an ethics of care. *Computers and Composition, 40,* 19–31. https://doi.org/10.1016/j.compcom.2016.03.006.

Colton, J. S., & Holmes, S. (2018). A social justice theory of active equality for technical communication. *Journal of Technical Writing and Communication, 48*(1), 4–30. https://doi.org/10.1177/0047281616647803

Colton, J.S., Holmes, S., & Walwema, J. (2017). From NoobGuides to #OpKKK: Ethics of Anonymous' tactical technical communication. *Technical Communication Quarterly, 26*(1), 59–75. https://doi.org/10.1080/10572252.2016.1257743

Company: About us. (2018). RealPage. https://www.realpage.com/company/

Coogan, D. (2002). Public rhetoric and public safety at the Chicago Transit Authority: Three approaches to accident analysis. *Journal of Business and Technical Communication, 16*(3), 277–305. https://doi.org/10.1177/1050651902016003002

Coppola, N., & Karis, B. (2000). *Technical communication, deliberative rhetoric, and environmental discourse: connections and directions.* Ablex Publishing.

Crow, A. (2013). Managing datacloud decisions and "big data": Understanding privacy choices in terms of surveillant assemblages. In H. A. McKee & D. N. DeVoss (Eds.), *Digital writing assessment & evaluation.* Computers and Composition Digital Press/Utah State University Press. http://ccdigitalpress.org/dwae/02_crow.html.

Crystal, A. (2007). Facets are fundamental: Rethinking information architecture frameworks. *Technical Communication, 54*(1), 16–26. https://www.jstor.org/stable/43089465

C-SPAN. (2013, June 18). *National Security Agency data collection programs.* C-SPAN. https://www.c-span.org/video/?313429-1/nsa-chief-testifies-damage-surveillance-leaks&start=2467

Cuijpers, C., & Koops, B. J. (2012). Smart Metering and Privacy in Europe: Lessons from the Dutch Case. In *European Data Protection: Coming of Age* (pp. 269–293). Springer. https://doi.org/10.1007/978-94-007-5170-5_12

Cummings, R. E. (2006). Coding with power: Toward a rhetoric of computer coding and composition. *Computers and Composition, 23*(4), 430–443. https://doi.org/10.1016/j.compcom.2006.08.002

Darmohray, T. (2010). *Job descriptions for system administrators* (3rd ed.). USENIX Assoc.

Data Justice Lab. (2021, June 11). *Data Justice 2021.* https://datajusticelab.org/data-justice-2021

Dautermann, J. (2005). Teaching business and technical writing in China: Confronting assumptions and practices at home and abroad. *Technical Communication Quarterly, 14*(2), 141–159. https://doi.org/10.1207/s15427625tcq1402_2

Day, M. (2000). Teachers at the crossroads: Evaluating teaching in electronic environments. *Computers and Composition, 17*, 31–40. https://doi.org/10.1016/S8755-4615(99)00028-6

De Angelis, M. (2007). *The beginning of history: Value struggles and global capital.* Pluto Press.

De Certeau, M. (1984). *The practice of everyday life* (S. Rendall, Trans.). University of California Press.

De Graaf, G. (2010). A report on reporting: Why peers report integrity and law violations in public organizations. *Public Administration Review, 70*(5), 767–779. https://doi.org/10.1111/j.1540-6210.2010.02204.x

De Hertogh, L. (2018). Feminist digital research methodology for rhetoricians of health and medicine. *Journal of Business and Technical Communication, 32*(4), 480–503. https://doi.org/10.1177/1050651918780188

Deleuze, G. (1992). Postscript on the societies of control. *October, 59*, 1–7.

Delgado, R., & Stefancic, J. (2012) *Critical race theory: An introduction.* New York University Press.

Dencik, L., Hintz, A., & Cable, J. (2019) Towards Data Justice: Bridging anti-surveillance and social justice activism. In D. Bigo, E. Isin, & E. Ruppert (Eds.), *Data politics: Worlds, subjects, rights* (pp. 167–186). Routledge.

Dencik, L., Hintz, A., & Carey, Z. (2017). Prediction, pre-emption and limits to dissent: Social media and big data uses for policing protests in the United Kingdom. *New Media & Society, 20*(4), 1433–1450. https://doi.org/10.1177/1461444817697722

DeVoss, D.N., Cushman, E. & Grabil, J. T. (2005). Infrastructure and composing: The when of new-media writing. *College Composition and Communication, 57*(1), 14–44.

DeWinter, J., & Vie, S. (2016). Games in technical communication. *Technical Communication Quarterly, 25*(3), 151–154. https://doi.org/10.1080/10572252.2016.1183411

Digital Rhetorical Privacy Collective. (n.d.). Digital Rhetorical Privacy Collective. Retrieved June 6, 2022, from https://drpcollective.com/

Ding, H. (2009). Rhetorics of alternative media in an emerging epidemic: SARS, censorship, and extra-institutional risk communication. *Technical Communication Quarterly, 18*(4), 327–350. https://doi.org/10.1080/10572250903149548

Dombrowski, P. (2000). *Ethics in technical communication*. Pearson.

Dombrowski, P. M. (2007). The evolving face of ethics in technical and professional communication: Challenger to Columbia. *IEEE Transactions on Professional Communication, 50*(4), 306–319. https://doi.org/10.1109/TPC.2007.908729

Donaldson, A. (2012). Surveillance and non-humans. In K. Ball, K. D. Haggerty, and D. Lyon (Eds.), *Routledge handbook of surveillance studies* (pp. 217–224). Routledge.

Dragga, S. (1997). A question of ethics: Lessons from technical communicators on the job. *Technical Communication Quarterly, 6*(2), 161–178. https://doi.org/10.1207/s15427625tcq0602_3

Dragga, S. (1999). Ethical intercultural technical communication: Looking through the lens of Confucian ethics. *Technical Communication Quarterly, 8*(4), 365–381, https://doi.org/10.1080/10572259909364675

Drew, C., & Sengupta, S. (2013, June 23). N.S.A. leak puts focus on system administrators. *The New York Times*. https://www.nytimes.com/2013/06/24/technology/nsa-leak-puts-focus-on-system-administrators.html

Drew, C., & Shane, S. (2013, July 4). Résumé shows Snowden honed hacking skills. *The New York Times*. https://www.nytimes.com/2013/07/05/us/resume-shows-snowden-honed-hacking-skills.html

Dubrofsky, R. E., & Magnet, S. A. (2015). *Feminist surveillance studies*. Duke University Press.

Duin, A., & Tham, J. (2020). The current state of analytics: Implications for learning management system (LMS) use in writing pedagogy. *Computers and Composition, 55*, 1–23. https://doi.org/10.1016/j.compcom.2020.102544

Duncan, M., & Hill, J. (2014). Termination documentation. *Business and Professional Communication Quarterly, 77*(3), 297–311. https://doi.org/10.1177/2329490614538806

D'urso, S. (2006). Who's watching us at work? Toward a structural-perceptual model of electronic monitoring and surveillance in organizations. *Communication Theory, 16*(3), 281–303. https://doi.org/10.1111/j.1468-2885.2006.00271.x

Dylko, I., Dolgov, I., Hoffman, W., Eckhart, N., Molina, M., & Aaziz, O. (2017). The dark side of technology: An experimental investigation of the influence of customizability technology on online political selective exposure. *Computers in Human Behavior, 73*, 181–190. https://doi.org/10.1016/j.chb.2017.03.031

Easter, B. (2018). "Feminist_brevity_in_light_of_masculine_long-windedness": code, space, and online misogyny. *Feminist Media Studies, 18*(4), 675–685. https://doi.org/10.1080/14680777.2018.1447335

Ede, L. (2004). *Situating composition: Composition studies and the politics of location.* Southern Illinois University Press.

Edenfield, A., Holmes, S., & Colton, J. (2019). Queering tactical technical communication: DIY HRT. *Technical Communication Quarterly, 28*(3), 177–191. https://doi.org/10.1080/10572252.2019.1607906

Ehrenfreund, M. (2013, Aug. 1). NSA surveillance scrutinized as Edward Snowden enters Russia. *The Washington Post.* https://www.washingtonpost.com/world/national-security/nsa-surveillance-scrutinized-as-edward-snowden-enters-russia/2013/08/01/c8ceb5d4-fad6-11e2-9bde-7ddaa186b751_story.html

Electronic Frontier Foundation. (n.d.). *Surveillance and human rights.* Electronic Frontier Foundation. https://www.eff.org/issues/surveillance-human-rights

Elnahla, N. (2020). *Black Mirror: Bandersnatch* and how Netflix manipulates us, the new gods. *Consumption Markets & Culture, 23*(5), 506–511. https://doi.org/10.1080/10253866.2019.1653288

Ellis, D., Tucker, I., Harper, D. (2013). The affective atmospheres of surveillance. *Theory & Psychology, 23*(6), 716–731. http://doi.org/10.1177/0959354313496604

Etezadzadeh, C. (2016). *Smart city—Future city? Smart city 2.0 as a livable city and future market.* Springer.

Esposito, R., & Cole, M. (2013, Aug. 26). How Snowden did it. NBC News. http://investigations.nbcnews.com/_news/2013/08/26/20197183-how-snowden-did-it?lite

European Data Protection Supervisor. (2020, June 24). *Homepage.* https://edps.europa.eu

Executive Order No. 10450, 3 C.F.R. 18 FR 2489 (1953).

Executive Order No. 12333. 3 C.F.R. 46 FR 59941 (1981).

Faber, B. D. (2001). Gen/ethics? Organizational ethics and student and instructor conflicts in workplace training. *Technical Communication Quarterly, 10*(3), 291–318. https://doi.org/10.1207/s15427625tcq1003_4

Fairweather, N. B. (1999). Surveillance in employment: The case of teleworking. *Journal of Business Ethics, 22,* 39–49. https://doi.org/10.1023/a:1006104017646

Farrier, J. (2007). The Patriot Act's institutional story: More evidence of congressional ambivalence. *PS: Political Science Politics, 40*(1), 93–97. https://doi.org/10.1017/S1049096507070151

Fauci, J., & Goodman, L. (2019). "You don't need nobody else knocking you down": Survivor-mothers' experiences of surveillance in domestic violence shelters. *Journal of Family Violence, 35*(3), 241–254. https://doi.org/10.1007/s10896-019-00090-y

Federal Bureau of Investigation. (2020, Jan. 15). *FOIA/PA overviews, exemptions, and terms.* https://www.fbi.gov/services/information-management/foipa/foia-pa-overviews-exemptions-and-terms

Fernandez, L. A., & Huey, L. (2009). Is resistance futile? Thoughts on resisting surveillance. *Surveillance & Society, 6*(3), 198–202. https://doi.org/10.24908/ss.v6i3.3280

Fielding, H. (2016). "Any time, any place": The myth of universal access and the semiprivate space of online education. *Computers and Composition, 40,* 103–114. https://doi.org/10.1016/j.compcom.2016.03.002.

Finn, J. (2012). Seeing surveillantly: Surveillance as social practice. In A. Doyle, R. Lippert, & D. Lyon (Eds.), *Eyes everywhere: The global growth of camera surveillance* (pp. 67–80). Routledge.

Fischer, J. (2004). Social responsibility and ethics: Clarifying the concepts. *Journal of Business Ethics, 52,* 381–390. https://doi-org.eur.idm.oclc.org/10.1007/s10551-004-2545-y

Foucault, M. (1977). *Discipline and punish: The birth of the prison*. (A. Sheridan, Trans.). Vintage.

Frith, J. (2019). *A billion little pieces: RFID and infrastructures of identification*. MIT Press.

Frost, E. (2020). Ultrasound, gender, and consent: An apparent feminist analysis of medical imaging rhetorics. *Technical Communication Quarterly, 30*(1), 1–15. https://doi.org/10.1080/10572252.2020.1774658

Frost, E., & Haas, A. (2017). Seeing and knowing the womb: A technofeminist reframing of fetal ultrasound toward a decolonization of our bodies. *Computers and Composition, 43,* 88–105. https://doi.org/10.1016/j.compcom.2016.11.004

Fuchs, C. (2011). New media, web 2.0 and surveillance: Web 2.0 surveillance. *Sociology Compass, 5*(2), 134–147. https://doi.org/10.1111/j.1751-9020.2010.00354.x

Fuchs, C. (2014). *Social media: A critical introduction*. Sage.

Fuchs, C., & Trottier, D. (2017). Internet surveillance after Snowden. *Journal of Information, Communication and Ethics in Society, 15*(4), 412–444. http://doi.org/10.1108/JICES-01-2016-0004

Gabbatt, A. (2014, Feb. 11). Protesters rally for "the day we fight back" against mass surveillance. *The Guardian*. https://www.theguardian.com/world/2014/feb/11/day-fight-back-protest-nsa-mass-surveillance

Galič, M., Timan, T., & Koops, B. (2017). Bentham, Deleuze and Beyond: An overview of surveillance theories from the panopticon to participation. *Philosophy & Technology, 30*(1), 9–37. https://doi.org/10.1007/s13347-016-0219-1

Gallagher, J. (2017). Writing for algorithmic audiences. *Computers and Composition, 45,* 25–35. https://doi.org/10.1016/j.compcom.2017.06.002

Gallagher, J., & Holmes, S. (2019). Empty templates: The ethical habits of empty state pages. *Technical Communication Quarterly, 28*(3), 271–283. https://doi.org/10.1080/10572252.2018.1564367

Gallie, W. B. (1956). Essentially contested concepts. *Proceedings of the Aristotelian Society, 1955–1956, 56*(1), 167–198. https://doi.org/10.1093/aristotelian/56.1.167

Gallucci, N. (2017, May 15). We need to talk about all these absurd stock photos of hackers. Mashable. https://mashable.com/article/horrible-hacker-stock-photos

Gandy, O. (1993). *The panoptic sort: A political economy of personal information*. Westview Press.

Gandy, O. (2011). Consumer protection in cyberspace. *Triple C, 9*(2), 175–189.

Gellman, B., & Markon, J. (2013, June 10). Edward Snowden says motive behind leaks was to expose 'surveillance state.' *The Washington Post*. https://www.washingtonpost.com/politics/edward-snowden-says-motive-behind-leaks-was-to-expose-surveillance-state/2013/06/09/aa3f0804-d13b-11e2-a73e-826d299ff459_story.html?noredirect=on

Gellman, B., and Poitras, L. (2013, June 7). U.S., British intelligence mining data from nine U.S. Internet companies in broad secret program. *The Washington Post*. https://www.washingtonpost.com/investigations/us-intelligence-mining-data-from-nine-us-internet-companies-in-broad-secret-program/2013/06/06/3a0c0da8-cebf-11e2-8845-d970ccb04497_story.html

Gibbons, M. (2018). The recalcitrant invention of X-ray images. *Technical Communication Quarterly, 28*(1), 54–68. https://doi.org/10.1080/10572252.2018.1539193

Gill, S. (1995). The global Panopticon? The neoliberal state, economic life, and democratic surveillance. *Alternatives: Global, Local, Political, 20*(1), 1–49. https://doi.org/10.1177/030437549502000101

Gilliom, J. (2001). *Overseers of the poor: Surveillance, resistances and the limits of privacy*. University of Chicago Press.

Gilliom, J., & Monahan, T. (2012). Everyday resistance. In D. Lyon, K. D. Haggerty, & K. Ball (Eds.), *Routledge handbook of surveillance studies* (pp. 405–411). Routledge.

Gjelten, T. (2013, Sept. 18). *Officials: Edward Snowden's leaks were masked by job duties*. NPR. https://www.npr.org/2013/09/18/223523622/officials-edward-snowdens-leaks-were-masked-by-job-duties

Goggin, P. N. (2013). *Environmental rhetoric and ecologies of place*. Routledge.

Gonzales, L., & DeVoss, D. N. (2016). Digging into data: Professional writers as data users. *Kairos: A Journal of Rhetoric, Technology, and Pedagogy, 20*(2). http://kairos.technorhetoric.net/20.2/topoi/beck-et-al/gon_devo.html.

Goode, L., & Godhe, M. (2017). Beyond capitalist realism—why we need critical future studies. *Culture Unbound, 9*(1). https://doi.org/10.3384/cu.2000.1525.1790615#sthash.7bZEHXNK.dpuf

Green, D. (2016). *How change happens*. Oxford University Press.

Greenwald, G. (2013a, June 17). Edward Snowden: NSA whistleblower answers reader questions. *The Guardian*. https://www.theguardian.com/world/2013/jun/17/edward-snowden-nsa-files-whistleblower

Greenwald, G. (2013b, July 31). XKeyscore: NSA tool collects 'nearly everything a user does on the internet.' *The Guardian*. https://www.theguardian.com/world/2013/jul/31/nsa-top-secret-program-online-data

Greenwald, G., MacAskill, E., & Poitras, L. (2013). Edward Snowden: The whistleblower behind the NSA surveillance revelations. *The Guardian*. https://www.theguardian.com/world/2013/jun/09/edward-snowden-nsa-whistleblower-surveillance

Haaksma, T., De Jong, M., & Karreman, J. (2018). Users' personal conceptions of usability and user experience of electronic and software products. *IEEE Transactions on Professional Communication*, *61*(2), 116–132. https://doi.org/10.1109/TPC.2018.2795398

Haas, A. (2012). Race, rhetoric, and technology: A case study of decolonial technical communication theory, methodology, and pedagogy. *Journal of Business and Technical Communication*, *26*(3), 277–310. https://doi.org/10.1177/1050651912439539

Haefner, J. (1999). The politics of the code. *Computers and Composition*, *16*(3), 325–339. https://doi.org/10.1016/S8755-4615(99)00014-6

Haggerty, K. D. (2006). Tear down the walls: On demolishing the Panopticon. In D. Lyon (Ed.), *Theorizing surveillance: The Panopticon and beyond* (pp. 23–45). Routledge.

Haggerty, K. D., & Ericson, R. (2000). The surveillant assemblage. *British Journal of Sociology*, *51*(4), 605–622.

Hall, L., & Wahlin, L. (2016). *A guide to technical communications: Strategies & applications*. Ohio State University Press. https://ohiostate.pressbooks.pub/engrtechcomm

Harding, L. (2014, Feb. 1). How Edward Snowden went from loyal NSA contractor to whistleblower. *The Guardian*. https://www.theguardian.com/world/2014/feb/01/edward-snowden-intelligence-leak-nsa-contractor-extract

Harner, S., & Rich, A. (2005). Trends in undergraduate curriculum in scientific and technical communication programs. *Technical Communication*, *52*(2), 209–220. https://www.jstor.org/stable/43089200

Hart-Davidson, W., Bernhardt, G., Mcleod, M., Rife, M., & Grabill, J. (2007). Coming to content management: Inventing infrastructure for organizational knowledge work. *Technical Communication Quarterly*, *17*(1), 10–34.

Harwell, D. (2020). Managers turn to surveillance software, always-on webcams to ensure employees are (really) working from home. *The Washington Post*. https://www.washingtonpost.com/technology/2020/04/30/work-from-home-surveillance

Hatch, O. G. (2018, Jan. 19). Text: S.139—115th Congress (2017–2018): FISA Amendments Reauthorization Act of 2017. https://www.congress.gov/bill/115th-congress/senate-bill/139/text/eah

Hausman, B. (2000). Rational management: Medical authority and ideological conflict in Ruth Lawrence's breastfeeding: A guide for the medical profession. *Technical Communication Quarterly: Medical Rhetoric, 9*(3), 271–289. https://doi.org/10.1080/10572250009364700

Hawisher, G. E., & Selfe, C. L. (1991). The rhetoric of technology and the electronic writing class. *College Composition and Communication, 42*(1), 55–65. https://doi.org//10.2307/357539

Hawkes, L. (2007). Impact of invasive web technologies on digital research. In H. A. McKee & D. N. DeVoss (Eds.), *Digital writing research: Technologies, methodologies, and ethical Issues* (pp. 337–351). Hampton Press.

Hayden, M. V. (2000, Apr. 12). *Statement for the record by Lt Gen Michael V. Hayden, USAF, director before the House Permanent Select Committee on Intelligence.* National Security Agency Central Security Service. https://www.nsa.gov/news-features/speeches-testimonies/Article/1620510/statement-for-the-record-by-lt-gen-michael-v-hayden-usaf-director-before-the-ho/

Hayes, B. (2012). The surveillance-industrial complex. In K. Ball, K. D. Haggerty, and D. Lyon (Eds.), *Routledge handbook of surveillance studies* (pp. 167–175). Routledge.

Healy, D. (1995). From place to space: Perceptual and administrative issues in the online writing center. *Computers and Composition, 12*, 183–93. https://doi.org/10.1016/8755-4615(95)90006-3

Henning, T., & Bemer, A. (2016). Reconsidering power and legitimacy in technical communication: A case for enlarging the definition of technical communicator. *Journal of Technical Writing and Communication, 46*(3), 311–341. http://doi.org/10.1177/0047281616639484

Henschel, S., & Meloncon, L. (2014). Of horsemen and layered literacies: Assessment instruments for aligning technical and professional communication undergraduate curricula with professional expectations. *Programmatic Perspectives, 6*(1), 3–26.

Hier, S. P., & Greenberg, J. (2007). *The surveillance studies reader.* Open University Press.

Hoebel, E. A. (1966). *Anthropology: The study of man* (3rd ed.). McGraw-Hill.

Holladay, D. (2017). Classified conversations: Psychiatry and tactical technical communication in Online Spaces. *Technical Communication Quarterly, 26*(1), 8–24.

Holwell, S. (2011). Fundamentals of information: Purposeful activity, meaning and conceptualisation. In M. Ramage & D. Chapman (Eds.), *Perspectives on information* (pp. 65–76). Routledge.

Hope, L. C. (2020). *Civility in the age of algorithms.* [Doctoral dissertation, Washington State University]. ProQuest Dissertations Publishing.

Hu, M. (2015). Taxonomy of the Snowden disclosures. *Washington and Lee Law Review, 72*(4), 1679–1768. https://ssrn.com/abstract=2730245

Hutchinson, L., & Novotny, M. (2018). Teaching a critical digital literacy of wearables: A feminist surveillance as care pedagogy. *Computers and Composition*, 50, 105–120. https://doi.org/10.1016/j.compcom.2018.07.006

Indeed. (2018, May 17). Technical writer II. Indeed. https://www.indeed.com/viewjob?jk=4dbaebb0da818a09&q=Technical+Intelligence+Analyst+Writer&tk=1ce48f2lib0cheu9&from=web&vjs=3

Ingraham, C., & Rowland, A. (2016). Performing imperceptibility: Google Street View and the tableau vivant. *Surveillance & Society*, 14(2), 211–226. http://doi.org/10.24908/ss.v14i2.6013

Introna, L. D. (2003). Workplace surveillance 'is' unethical and unfair. *Surveillance & Society*, 1(2), 210–216. www.doi.org/10.24908/ss.v1i2.3354

Jackson, S. (2002, Nov. 18). Beyond the fatherless homes. *The Kansas City Star*. https://sites.google.com/view/sarahyoungphd/publications/newspaper/the-kansas-city-star?authuser=0

Jaeger, P. T., & Burnett, G. (2010). *Information worlds: Social context, technology, and information behaviour in the age of the internet*. Routledge.

Janangelo, J. (1991). Technopower and technoppression: Some abuses of power and control in computer-assisted writing environments. *Computers and Composition*, 9(1), 47–64. https://doi.org/10.1016/8755-4615(91)80038-F

Jensen, O. B. (2016). New 'Foucauldian boomerangs': Drones and urban surveillance. *Surveillance & Society*, 14(1), 20–33.

Johnson, G. P. (2020). Grades as technology of surveillance: Normalization, control, and big data in the teaching of writing. In E. Beck & L. Hutchinson Campos (Eds.), *Privacy matters: Conversations about surveillance within and beyond the classroom* (53–72). University Press of Colorado.

Johnson-Eilola, J., & Selber, S. (2013). *Solving problems in technical communication*. University of Chicago Press.

Johnson-Sheehan, R. (2002). *Writing proposals: Rhetoric for managing change*. Pearson Education.

Johnson-Sheehan, R., & Morgan, L. (2009). Conservation Writing: An Emerging Field in Technical Communication. *Technical Communication Quarterly*, 18(1), 9–27. https://doi.org/10.1080/10572250802437283

Jones, J. (2015). Information graphics and intuition. *Journal of Business and Technical Communication*, 29(3), 284–313. https://doi.org/10.1177/1050651915573943

Jones, N. N. (2016). The technical communicator as advocate: Integrating a social justice approach in technical communication. *Journal of Technical Writing and Communication*, 46(3), 342–361. https://doi.org/10.1177/0047281616639472

Jones, N. N. Moore, K. R., & Walton, R. (2016). Disrupting the past to disrupt the future: An antenarrative of technical communication. *Technical Communication Quarterly*, 25(4), 211–229. https://doi.org/10.1080/10572252.2016.1224655

Kabel, S. C., Wielinga, B. J., & de Hoog, R. (2000). *Ontologies for indexing technical manuals for instruction*. TNO.

Kalmbach, J. (2007). Technical reports as rhetorical practice. In C. L. Selfe (Ed.), *Resources in technical communication: Outcomes and approaches* (21–35). Baywood Publishing.

Kammerer, D. (2012). Surveillance in literature, film and television. In D. Lyon, K.D. Haggerty, & K. Ball (Eds.), *Routledge handbook of surveillance studies* (pp. 99–106). Routledge.

Kegu, J. (2019, Sept. 16). *Edward Snowden wants to come home: "I'm not asking for a pass. What I'm asking for is a fair trial."* CBS News. https://www.cbsnews.com/news/edward-snowden-nsa-cbs-this-morning-interview-today-2019-09-16/

Ken, I. (2010). Race, Class, and Gender as Organizing Principles. In *Digesting Race, Class, and Gender* (pp. 15–51). Palgrave Macmillan US. https://doi.org/10.1057/9780230115385_2

Kessler, M., & Graham, S. (2018). Terminal node problems: ANT 2.0 and prescription drug labels. *Technical Communication Quarterly, 27*(2), 121–136. https://doi.org/10.1080/10572252.2018.1425482

Killingsworth, M. J. (2005). From Environmental rhetoric to ecocomposition and ecopoetics: Finding a place for professional communication. *Technical Communication Quarterly, 14*(4), 359–373. https://doi.org/10.1207/s15427625tcq1404_1

Kimball, M. (2005). Database e-portfolio systems: A critical appraisal. *Computers and Composition, 22*, 434–58. https://doi.org/10.1016/j.compcom.2005.08.003

Kimball, M. (2006). Cars, culture, and tactical technical communication. *Technical Communication Quarterly, 15*(1), 67–86.

Kimball, M. (2017a). The golden age of technical communication. *Journal of Technical Writing and Communication, 47*(3), 330–358.

Kimball, M. (2017b). Tactical technical communication. *Technical Communication Quarterly, 26*(1), 1–7.

Kmiec, D., & Longo, B. (2017). *The IEEE guide to writing in the engineering and technical fields*. John Wiley & Sons.

Knievel, M. (2008). Police reform, task force rhetoric, and traces of dissent: Rethinking consensus-as-outcome in collaborative writing situations. *Journal of Technical Writing and Communication, 38*(4), 331–362. https://doi.org/10.2190/TW.38.4.c

Koops, B., Newell, B., Timan, T., Skorvanek, I., Chokrevski, T., & Galic, M. (2017). A typology of privacy. *University of Pennsylvania Journal of International Law, 38*(2), 483–575. https://scholarship.law.upenn.edu/jil/vol38/iss2/4/

Kostelnick, C. (2019). *Humanizing visual design*. Routledge.

Kreibich, R., Oertel, B., & Wolk, M. (2011). Futures studies and future-oriented technology analysis principles, methodology and research questions. *1st Berlin Symposium on Internet and Society, Das Alexander von Humboldt*

Institut für Internet und Gesellschaft. https://www.hiig.de/wp-content/uploads/2012/04/Foresight-Paper.pdf

Kupperman, J. J. (1969). Nuance and ethical choice. *Ethics, 79*(2), 105–114. http://www.jstor.org/stable/2379174

Lam, L. (2013, June 25). *Snowden sought Booz Allen job to gather evidence on NSA surveillance*. South China Morning Post. http://www.scmp.com/news/hong-kong/article/1268209/snowden-sought-booz-allen-job-gather-evidence-nsa-surveillance

Landau, S. (2013). Making sense from Snowden: What's significant in the NSA surveillance revelations. *IEEE Security and Privacy, 11*(4), 54–63.

Lauer, C., & Brumberger, E. (2016). Technical communication as user experience in a broadening industry landscape. *Technical Communication, 63*(3), 248–264.

Lawrence, H. Y., Fernandez, L., Lussos, R. G., Stabile, B., & Broeckelman-Post, M. (2019). Communicating campus sexual assault: A mixed methods rhetorical analysis. *Technical Communication Quarterly, 28*(4), 299–316. https://doi.org/10.1080/10572252.2019.1621386

Lee, D. (2022). The ethics of extrapolation: Science fiction in the technical communication classroom. *Technical Communication Quarterly, 31*(1), 77–88. https://doi.org/10.1080/10572252.2020.1866678

Li, L. (2020). Visualizing Chinese immigrants in the U.S. statistical atlases: A case study in charting and mapping the other(s). *Technical Communication Quarterly, 29*(1), 1–17. https://doi.org/10.1080/10572252.2019.1690695

Liberty. (n.d.). Merriam-Webster. https://www.merriam-webster.com/dictionary/liberty

Library of Congress. (2017, Feb. 24). *Engineering disciplines*. United States Library of Congress. https://www.loc.gov/rr/scitech/SciRefGuides/eng-disciplines.html

Lindgren, C. A. (2021). Writing with data: A study of coding on a data-journalism team. *Written Communication, 38*(1), 114–162. https://doi.org/10.1177/0741088320968061

Litt, E. (2012). Knock, knock. Who's there? The imagined audience. *Journal of Broadcasting & Electronic Media, 56*, 330–345.

Lockett, A. L. (2013). Leaked: A grammar of information in surveillance cultures (Order No. 29267279). Available from ProQuest Dissertations & Theses Global. (2672351193). https://www.proquest.com/dissertations-theses/leaked-grammar-information-surveillance-cultures/docview/2672351193/se-2?accountid=13598

Loi, Hauser, C., & Christen, M. (2020). Highway to (digital) surveillance: When are clients coerced to share their data with insurers? *Journal of Business Ethics, 175*(1), 7–19. https://doi.org/10.1007/s10551-020-04668-1

Lucas, G. R., Jr. (2014). NSA management directive #424: Secrecy and privacy in the aftermath of Edward Snowden. *Ethics & International Affairs, 28*(1), 29–38. https://doi.org/10.1017/S0892679413000488

Lyon, D. (1994). *The electronic eye: The rise of surveillance society*. Polity.
Lyon, D. (2001). *Surveillance society: Monitoring everyday life*. Open University Press.
Lyon, D. (2002). Everyday surveillance: Personal data and social classifications. *Information Communication & Society*, 5(2), 242–257. https://doi.org/10.1080/13691180210130806
Lyon, D. (2007). *Surveillance studies: An overview*. Polity Press.
Lyon, D. (2009). *Identifying citizens: ID cards as surveillance*. Polity Press.
Lyon, D. (2015). *Surveillance after Snowden*. Polity Press.
Lyon, D. (2017). Surveillance culture: Engagement, exposure, and ethics in digital modernity. *International Journal of Communication*, 11, 824–842. https://ijoc.org/index.php/ijoc/article/view/5527
Lyon, D. Haggerty, K. D., & Ball, K. (2012). Introducing surveillance studies. In D. Lyon, K. D. Haggerty, & K. Ball (Eds.), *Routledge handbook of surveillance studies* (pp. 1–11). Routledge.
MacAskill, E., Borger, J., Hopkins, N., Davies, N., & Ball, J. (2013, June 21). GCHQ taps fibre-optic cables for secret access to world's communications. *The Guardian*. https://www.theguardian.com/uk/2013/jun/21/gchq-cables-secret-world-communications-nsa
MacAskill, E., & Hern, A. (2018, June 4). Edward Snowden: 'The people are still powerless, but now they're aware.' *The Guardian*. https://www.theguardian.com/us-news/2018/jun/04/edward-snowden-people-still-powerless-but-aware
MacDonald-Evoy, J. (2017, Feb. 13). Stingray [Surveillance Technology Documentary]. YouTube. https://www.youtube.com/watch?v=Y1E3r2USd8I]
Machili, I., Angouri, J., & Harwood, N. (2019). "The snowball of emails we deal with": CCing in multinational companies. *Business and Professional Communication Quarterly*, 82(1), 5–37. https://doi.org/10.1177/2329490618815700
Macintosh, N. (2006). *Accounting, accountants, and accountability*. Routledge.
MacMillan, S. (2012). The promise of ecological inquiry in writing research. *Technical Communication Quarterly*, 21(4), 346–361. https://doi.org/10.1080/10572252.2012.674873
Malone, E., & Wright, D. (2018). To promote that demand: Toward a history of the marketing white paper as a genre. *Journal of Business and Technical Communication*, 32(1), 113–147. https://doi.org/10.1177/1050651917729861
Mann, S., Nolan, J., & Wellman, B. (2003). Sousveillance: Inventing and using wearable computing devices for data collection in surveillance environments. *Surveillance & Society*, 1(3), 331–355. http://doi.org/10.24908/ss.v1i3.3344
Markel, M. (2001). *Ethics in technical communication: A critique and synthesis*. Ablex.
Markel, M. (2005). The rhetoric of misdirection in corporate privacy-policy statements, *Technical Communication Quarterly*, 14(2), 197–214. https://doi.org/10.1207/s15427625tcq1402_5
Markel, M. (2009a). Anti-employer blogging: An overview of legal and ethical issues. *Journal of Technical Writing and Communication*, 39(2), 123–139. https://doi.org/10.2190/TW.39.2.b

Markel, M. (2009b). Time and exigence in temporal genres. *Journal of Business and Technical Communication, 23*(1), 3–27. https://doi.org/10.1177/1050651908324376

Marsh, B. (2004). Turnitin.com and the scriptural enterprise of plagiarism detection. *Computers and Composition, 21*, 427–438. https://doi.org/10.1016/j.compcom.2004.08.002

Marwick, A, & boyd, d. (2014). Networked privacy: How teenagers negotiate context in social media. *New Media & Society, 16*(7), 1051–1067. https://doi.org/10.1177/1461444814543995

Marx, G. (1998). Ethics for the new surveillance. *The Information Society, 14*(3), 171–185. https://doi.org/10.1080/019722498128809

Marx, G. T. (2003). A tack in the shoe: Neutralizing and resisting the new surveillance. *Journal of Social Issues, 59*(2), 369–390. https://doi.org/10.1111/1540-4560.00069

Marx, G. T. (2005). Surveillance and society. In G. Ritzer (Ed.), *Encyclopedia of social theory* (Vol. 1, pp. 817–821). SAGE. https://doi.org/10.4135/9781412952552.n303

Marx, G. T. (2012). Preface. In D. Lyon, K. D. Haggerty, & K. Ball (Eds.), *Routledge handbook of surveillance studies* (pp. xx–xxxi). Routledge.

Marx, G. T. (2016). *Windows into the soul: Surveillance and society in an age of high technology*. University of Chicago Press.

Masur, P. K., & Scharkow, M. (2016). Disclosure management on social network sites: Individual privacy perceptions and user-directed privacy strategies. *Social Media + Society, 2*(1), 1–13. https://doi.org/10.1177/2056305116634368

McCarthy, T. (2013, June 9). Edward Snowden identifies himself as source of NSA leaks—as it happened. *The Guardian*. https://www.theguardian.com/world/2013/jun/09/nsa-secret-surveillance-lawmakers-live

McCulloch, E., Shiri, A., & Nicholson, D. (2005). Challenges and issues in terminology mapping: A digital library perspective. *The Electronic Library, 23*(6), 671–677. https://doi.org/10.1108/02640470510635755

McGrath, J. (2012). Performing surveillance. In K. Ball, K. Haggerty, and D. Lyon (Eds.), *Routledge handbook of surveillance studies* (pp. 83–90). Routledge.

McCool, M. (2006). Information architecture: Intercultural human factors. *Technical Communication, 53*(2), 167–183.

McKee, H. A. (2011). Policy matters now and in the future: Net neutrality, corporate data mining, and government surveillance. *Computers and Composition, 28*(4), 276–291. https://doi.org/10.1016/j.compcom.2011.09.001

McKee, H. A. (2016). Protecting net neutrality and the infrastructure of internet delivery: Considerations for our past, present, and future. *Kairos: A Journal of Rhetoric, Technology, and Pedagogy, 20*(2). http://kairos.technorhetoric.net/20.2/topoi/beck-et-al/mckee.html

McKee, H. A. & Porter, J. E. (2010). Legal and regulatory issues for technical communicators conducting global internet research. *Technical Communication, 57*(3), 282–299.

McQuade, S. C., & Danielson, P. (2005). Monitoring and surveillance. In C. Mitcham (Ed.), *Encyclopedia of Science, Technology, and Ethics* (pp. 1228–1232). Macmillan Reference.

Messina, C. (2021). The critical fan toolkit: Fanfiction genres, ideologies, and pedagogies. [Doctoral dissertation, Northeastern University]. ProQuest Dissertations Publishing.

Meyer, M. (1996). Rhetoric and the theory of argument. *Revue Internationale de Philosophie, 50*(196), 325–357. http://www.jstor.org/stable/23954812

Mills, C. (1997). *The racial contract*. Cornell University Press.

Mirel, B., Barton, E., & Ackerman, M. (2008). Researching telemedicine: Capturing complex clinical interactions with a simple interface design. *Technical Communication Quarterly, 17*(3), 358–378. https://doi.org/10.1080/10572250802100477

Moeller, M., & Frost, E. (2016). Food fights: Cookbook rhetorics, monolithic constructions of womanhood, and field narratives in technical communication. *Technical Communication Quarterly, 25*(1), 1–11. https://doi.org/10.1080/10572252.2016.1113025

Monahan, T. (2011). Surveillance as cultural practice. *The Sociological Quarterly, 52*(4), 495–508. http://www.jstor.org/stable/23027562

Monahan, T., & Murakami Wood, D. (2018). *Surveillance studies: A reader*. Oxford University Press.

Monahan, T., Phillips, D. J., Murakami Wood, D. (2010). Surveillance and empowerment. *Surveillance & Society, 8*(2), 106–112. https://doi.org/10.24908/ss.v8i2.3480

Moran, C. (1995). Notes toward a rhetoric of e-mail. *Computers and Composition, 12*, 15–21. https://doi.org/10.1016/8755-4615(95)90019-5

Morville, P. (2005). *Ambient findability*. O'Reilly.

Mulligan, D. K., Koopman, C., & Doty, N. (2016). Privacy is an essentially contested concept: A multi-dimensional analytic for mapping privacy. *Philosophical Transactions of the Royal Society. A, 374*(2083). https://doi.org/10.1098/rsta.2016.0118

Murakami Wood, D., & Wright, S. (2015). Before and after Snowden. *Surveillance & Society, 13*(2), 132–138. https://doi.org/10.24908/ss.v13i2.5710

Myre, G. (2019, Dec. 14). Stories of the decade: Edward Snowden and mass surveillance. NPR. https://www.npr.org/2019/12/14/787891267/stories-of-the-decade-edward-snowden-and-mass-surveillance

National Security Agency. (n.d.a). *Mission and Values*. https://www.nsa.gov/about/mission-values/

National Security Agency. (n.d.b). *NSA careers*. https://www.intelligencecareers.gov/nsa/

Near, J. P., & Miceli, M. P. (1996). Whistle-blowing: Myth and reality. *Journal of Management, 22*(3), 507–526.

New York Times. (2020). Times topics: Surveillance of citizens by government. https://www.nytimes.com/topic/subject/surveillance-of-citizens-by-government

Nissenbaum, H. (2009). *Privacy in context: Technology, policy, and the integrity of social life*. Stanford Law Books.

Novotny, M., & Hutchinson, L. (2019). Data our bodies tell: Towards critical feminist action in fertility and period tracking applications. *Technical Communication Quarterly, 28*(4), 332–360. https://doi.org/10.1080/10572252.2019.1607907

NSA/CSSM 1-52. (2012). *Fairview: Dataflow diagrams*. ProPublica. https://www.documentcloud.org/documents/2274318-fairviewdataflowchartsapril2012.html

Office of Policy Development and Research. (n.d.). *Housing, inclusion, and public safety*. HUD USER. Retrieved June 17, 2022, from https://www.huduser.gov/portal/periodicals/em/summer16/highlight1.html

Opel, D. (2017). Ethical research in "Health 2.0": Considerations for scholars of medical rhetoric. In J. B. Scott, & L. Meloncon (Eds.), *Methodologies for the rhetoric of health and medicine* (pp. 176–194). Routledge.

Opel, D., & Rhodes, J. (2018). Beyond student as user: Rhetoric, multimodality, and user-centered design. *Computers and Composition, 49*, 71–81. https://doi.org/10.1016/j.compcom.2018.05.008

Orwell, G. (1950). *1984*. Signet. (Original work published 1949).

Page, S. (2015, May 10). Debate over limits of government surveillance and the future of the Patriot Act. *The Diane Rehm Show*. https://dianerehm.org/shows/2015-05-11/debate-over-limits-of-government-surveillance-and-the-future-of-the-patriot-act

Parker, D. B. (1976). *Crime by computer*. Charles Scribner's Sons.

Parkinson, J. R. (2013, June 18). NSA: 'Over 50' terror plots foiled by data dragnets. *ABC News*. https://abcnews.go.com/Politics/nsa-director-50-potential-terrorist-attacks-thwarted-controversial/story?id=19428148#.UcGmZaLVDms

Peiser, J. (2022, June 14). Fans told Lizzo a word in her song was offensive. She changed the lyrics. *The Washington Post*. https://www.washingtonpost.com/nation/2022/06/14/lizzo-ableist-slur-lyric-apology/

Penney, J. (2021). Understanding chilling effects. *106 Minnesota Law Review, 101*, 101–191. https://ssrn.com/abstract=3855619

Penney, J. W. (2017). Internet surveillance, regulation, and chilling effects online: a comparative case study. *Internet Policy Review, 6*(2). https://doi.org/10.14763/2017.2.692

Perelman, C. (1979). The new rhetoric: A theory of practical reasoning. In *The new rhetoric and the humanities* (Vol. 140). Springer. https://doi.org/10.1007/978-94-009-9482-9_1

Petersen, E. (2019). The "reasonably bright girls": Accessing agency in the technical communication workplace through interactional power. *Technical*

Communication Quarterly, 28(1), 21–38. https://doi.org/10.1080/10572252.2018.1540724

Pflugfelder, E. (2017). Reddit's 'explain like I'm five': Technical descriptions in the wild. *Technical Communication Quarterly, 26*(1), 25–41. https://doi.org/10.1080/10572252.2016.1257741

Phelps, J. L. (2022). Concomitant Ethics: Institutional Review Boards and Technical and Professional Communication's Social Justice Turn. *Journal of Business and Technical Communication, 36*(3), 270–295. https://doi.org/10.1177/10506519221087709

Pigg, S. (2014). Coordinating constant invention: Social media's role in distributed work. *Technical Communication Quarterly, 23*(2), 69–87. https://doi.org/10.1080/10572252.2013.796545

Pihlaja, B. (2022). Everyday ethics at the border: Normative ethics for the 21st century. *Journal of Business and Technical Communication, 36*(3), 296–325. https://doi.org/10.1177/10506519221087937

Plesnicar, M. M., & Sarf, P. (2020). 'This web page should not exist': A case study of online shaming in Slovenia. In D. Trottier, R. Gabdulhakov, & Q. Huang (Eds.), *Introducing vigilant audiences* (pp. 187–214). Open Book Publishers.

Popular Mechanics. (2020). Homepage. *Popular Mechanics.* https://www.popularmechanics.com/

Popular Science. (2020). Homepage. *Popular Science.* https://www.popsci.com

Porat, M. U. (1977). *The information economy: Definition and measurement.* U.S. Government Printing Office. https://eric.ed.gov/?id=ED142205

Porat, M. U., & Rubin, M. R. (1967). *The information economy: The technology matrices.* U.S. Government Printing Office. https://books.google.nl/books?hl=en&lr=&id=MANPAAAAMAAJ&oi=fnd&pg=PP11&dq=information+economy&ots=2vnyVZpzHx&sig=G8pKNYJF_GXfnTNGm6atbK-RzmM&redir_esc=y#v=onepage&q=information%20economy&f=false

Porter, J. E. (1993). Developing a postmodern ethics of rhetoric and composition. In T. Enos, & S. C. Brown (Eds.), *Defining the new rhetorics* (pp. 207–226). Sage.

Porter, J. E. (2013). How can rhetoric theory inform the practice of technical communication? In J. Johnson-Eilola, & S. A. Selber (Eds.), *Solving problems in technical communication* (pp. 125–145). University of Chicago Press.

Powell, T. (2019). Ethics. In T. Reardon, T. Powell, J. Arnett, M. Logan, & C. Race (Eds.), *Open technical communication.* https://digitalcommons.kennesaw.edu/opentc/16

Pridmore, J. (2012). Consumer surveillance: Context, perspectives and concerns in the personal information economy. In D. Lyon, K. D. Haggerty, & K. Ball (Eds.), *Routledge handbook of surveillance studies* (pp. 321–329). Routledge.

Privacy International. (n.d.). *Mass surveillance.* Privacy International. https://privacyinternational.org/learn/mass-surveillance

Privacy Writer Jobs. (2020, June 10). Indeed.com. https://www.indeed.com/q-privacy-Writer-jobs.html?vjk=dc5f7c3b58c7a038

Purdy, J. (2009). Anxiety and the archive: Understanding plagiarism detection services as digital services. *College Composition and Communication*, 26, 65–77. https://doi.org/10.1016/j.compcom.2008.09.002

Raco, M. (2007). Securing sustainable communities: Citizenship, safety and sustainability in the new urban planning. *European Urban and Regional Studies*, 14(4), 305–320. https://doi.org/10.1177/0969776407081164

Ramage, M., & Chapman, D. (2011). Introduction. In M. Ramage, & D. Chapman (Eds.), *Perspectives on information* (pp. 1–7). Routledge.

Ramey, J. (2014). The coffee planter of Saint Domingo: A technical manual for the Caribbean slave owner. *Technical Communication Quarterly*, 23(2), 141–159. https://doi.org/10.1080/10572252.2013.811164

RAND Corporation (2020). *Science and Technology*. https://www.rand.org/topics/science-and-technology.html

Ranney, F. (2000). Beyond Foucault: Toward a user-centered approach to sexual harassment policy. *Technical Communication Quarterly*, 9(1), 9–28. https://doi.org/10.1080/10572250009364683

Reardon, D., Wright, D., & Malone, E. (2017). Quest for the happy ending to Mass Effect 3: The challenges of cocreation with consumers in a post-Certeauian age. *Technical Communication Quarterly*, 26(1), 42–58. https://doi.org/10.1080/10572252.2016.1257742

Rebbit. (2020, Feb. 28). *Casino surveillance tech shares casino secrets about casino (r/IAmA)*. YouTube. https://www.youtube.com/watch?v=bLVLIy5qHqM).

Reilly, C. A. (2016). Coming to terms: Critical approaches to ubiquitous digital surveillance. *Kairos: A Journal of Rhetoric, Technology, and Pedagogy*, 20(2). http://kairos.technorhetoric.net/20.2/topoi/beck-et-al/reilly.html#main.

Rest, J. (1986). *Moral development: Advances in research and theory*. Praeger.

Reuters. (2013, Oct. 26). *Protesters march in Washington against NSA spying*. Reuters. https://www.reuters.com/article/us-usa-security-protest-idUSBRE99P0B420131026

Reyman, J. (2013). User data on the social web: Authorship, agency, and appropriation. *College English*, 75(5), 513–33.

Rich, A. (1984). Notes toward a politics of location. *Blood, bread, and poetry: selected prose, 1979–1985* (pp. 210–231). W. W. Norton.

Richards, A., & David, C. (2005). Decorative color as a rhetorical enhancement on the world wide web. *Technical Communication Quarterly*, 14(1), 31–48. https://doi.org/10.1207/s15427625tcq1401_4

Richards, J. L., Lenart, J., Sumner, D., & Christensen, D. (2018). From big ag to campus cafeterias: Intersections of food-supply networks as technical communication pedagogy. *Open Library of Humanities*, 4(2). https://doi.org/10.16995/olh.381

Richards, N. M. (2013). The dangers of surveillance. *Harvard Law Review*, 126(7), 1934–1965.

Risen, J., & Poitras, L. (2013, Sept. 28). N.S.A. gathers data on social connections of U.S. citizens. *The New York Times*. https://www.nytimes.com/2013/09/29/us/nsa-examines-social-networks-of-us-citizens.html

Rose, E. J., Racadio, R., Wong, K., Nguyen, S., Kim, J., & Zahler, A. (2017). Community-based user experience: Evaluating the usability of health insurance information with immigrant patients. *IEEE Transactions on Professional Communication, 60*(2), 214–231. http://doi.org/10.1109/TPC.2017.2656698

Ruby, F., Goggin, G., & Keane, J. (2017) "Comparative silence" still? Journalism, academia, and the five eyes of Edward Snowden. *Digital Journalism, 5*(3), 353–367. http://doi.org/10.1080/21670811.2016.1254568

Rude, C. D. (1988). Format in Instruction Manuals: Applications of Existing Research. *Iowa State Journal of Business and Technical Communication, 2*(1), 63–77. https://doi.org/10.1177/105065198800200105

Salvo, M. J. (2001) Ethics of engagement: User-centered design and rhetorical methodology, *Technical Communication Quarterly, 10*(3), 273–290. https://doi.org/10.1207/s15427625tcq1003_3

Sánchez, F. (2020). Examining methectic technical communication in an urban planning comic book. *Technical Communication Quarterly, 29*(3), 287–303. https://doi.org/10.1080/10572252.2020.1768289

Sarat-St. Peter, H. (2017). "Make a bomb in the kitchen of your mom": Jihadist tactical technical communication and the everyday practice of cooking. *Technical Communication Quarterly: Tactical Technical Communication, 26*(1), 76–91. https://doi.org/10.1080/10572252.2016.1275862

Schneider, S. (2005). Usable pedagogies: Usability, rhetoric, and sociocultural pedagogy in the technical writing classroom. *Technical Communication Quarterly, 14*(4), 447–467. https://doi.org/10.1207/s15427625tcq1404_4

Scott, H., Fawkner, S., Oliver, C. W., & Murray, A. (2017). How to make an engaging infographic? *British Journal of Sports Medicine, 51*(16), 1183–1184. https://doi.org/10.1136/bjsports-2016-097023

Scott, R. L. (1967). On viewing rhetoric as epistemic. *Central States Speech Journal, 18*, 9–17.

Seawright, L. (2017). *Genre of power: Police report writers and readers in the justice system*. Conference on College Composition and Communication of the National Council of Teachers of English.

Seigel, M. (2013). *The rhetoric of pregnancy*. Chicago: University of Chicago Press.

Selfe, C. L. (Ed.). (2007). *Resources in technical communication: outcomes and approaches*. Routledge.

Selfe, C. L., & Selfe, Richard J., Jr. (1994). The politics of the interface: Power and its exercise in electronic contact zones. *College Composition and Communication, 45*(4), 480–504. https://doi.org/10.2307/358761

Sewell, G., & Barker, J. R. (2007). Neither good, nor bad, but dangerous: surveillance as an ethical paradox. In S. P. Hier and J. Greenberg (Eds.), *The surveillance studies reader*, 345–367.

Shane, S. (2013, June 21). Ex-contractor is charged in leaks on N.S.A. surveillance. *The New York Times*. https://www.nytimes.com/2013/06/22/us/snowden-espionage-act.html

Shane, S., & Sanger, D. E. (2013, June 30). Job title key to inner access held by Snowden. *The New York Times*. https://www.nytimes.com/2013/07/01/us/job-title-key-to-inner-access-held-by-snowden.html

Shaw, J. (2017). The watchers. *Harvard Magazine*. https://www.harvardmagazine.com/2017/01/the-watchers

Sherlock, L. (2009). Genre, activity, and collaborative work and play in World of Warcraft: places and problems of open systems in online gaming. *Journal of Business and Technical Communication*, 23(3), 263–293. https://doi.org/10.1177/1050651909333150

Siegel, R., & Johnson, C. (2013, June 21). *U.S. charges NSA leaker Snowden with espionage*. NPR. https://www.npr.org/2013/06/21/194371404/u-s-charges-nsa-leaker-snowden-with-espionage

SIGWROC. (n.d.). *Writing and Rhetoric of Code*. Retrieved June 14, 2022, from https://wroc.netlify.app/

Smith, G. J. D. (2012). Surveillance work(ers). In D. Lyon, K. D. Haggerty, & K. Ball (Eds.), *Routledge handbook of surveillance studies* (pp. 107–15). Routledge.

Snowden, E. (@Snowden). (n.d.). *Tweets* [Twitter profile]. https://twitter.com/Snowden

Snowden, E. (2019a). *Permanent record*. Macmillan.

Snowden, E. (2019b, Sept. 15). Edward Snowden on 9/11 and why he joined the army: 'Now, finally, there was a fight.' *The Guardian*. https://www.theguardian.com/us-news/2019/sep/15/edward-snowden-on-911-and-why-he-joined-the-army-now-finally-there-was-a-fight

Society for Technical Communication [STC]. (2018). *Defining technical communication*. STC. https://www.stc.org/about-stc/defining-technical-communication/

Solove, D. J. (2004). Reconstructing electronic surveillance law. *George Washington Law Review*, 72(6), 1264–1305. http://dx.doi.org/10.2139/ssrn.445180

Solove, D. J. (2011). Why privacy matters even if you have "nothing to hide." *The Chronicle of Higher Education*, 57(37), B11–B13. http://www.uvm.edu/~dguber/POLS21/articles/solove.htm

South China Morning Post. (2013, June 15). *Protesters urge Hong Kong to protect Snowden, demand answers on US spying*. https://www.scmp.com/news/hong-kong/article/1261433/protesters-urge-hong-kong-protect-snowden-demand-answers-us-spying

Spinuzzi, C. (2007). Guest editor's introduction: Technical communication in the age of distributed work. *Technical Communication Quarterly*, *16*(3), 265–277. https://doi.org/10.1080/10572250701290998

Staples, W. G. (2000). *Everyday surveillance: Vigilance and visibility in postmodern life* (2nd ed.). Rowman & Littlefield.

Stark, L. (2016). The emotional context of information privacy. *The Information Society*, *32*(1), 14–27. https://doi.org/10.1080/01972243.2015.1107167

Stoddart, E. (2012). "A surveillance of care." In K. Ball, K. D. Haggerty, and D. Lyon (Eds.), *Routledge handbook of surveillance studies* (pp. 369–376). Routledge.

Sullivan, P. A., & Porter, J. E. (1993). Remapping curricular geography: Professional writing in/and English. *Journal of Business and Technical Communication*, *7*(4), 389–422.

Surveillance Studies Network. (n.d.). *Welcome to the surveillance studies network*. SSN. https://www.surveillance-studies.net

Surveillance Writer Jobs. (2020, June 10). Indeed. https://www.indeed.com/q-surveillance-Writer-jobs.html

Teston, C. (2012). Moving from artifact to action: A grounded investigation of visual displays of evidence during medical deliberations. *Technical Communication Quarterly*, *21*(3), 187–209. https://doi.org/10.1080/10572252.2012.650621

Tham, J., & Duin, A. H. (2020). Digital literacy in an age of pervasive surveillance: A case of wearable technology. In E. Beck & E. H. Campos (Eds.), *Privacy matters: Conversations about surveillance within and beyond the Classroom* (pp. 93–112). University Press of Colorado.

Tham, J., Rosselot-Merritt, J., Veeramoothoo, S., Bollig, N., & Duin, A. (2020). Toward a radical collaboratory model for graduate research education: A collaborative autoethnography. *Technical Communication Quarterly*, *29*(4), 1–17. https://doi.org/10.1080/10572252.2020.1713404

Thatcher, B., St. Amant, K., & Sides, C. (2011). *Teaching intercultural rhetoric and technical communication*. Routledge. https://doi.org/10.4324/9781315223605

TikTok. (2020). Privacy policy. TikTok. https://www.tiktok.com/legal/privacy-policy?lang=en

Toomey, P., & Gorski, A. (2021, Sept. 7). The privacy lesson of 9/11: Mass surveillance is not the way forward. American Civil Liberties Union. https://www.aclu.org/news/national-security/the-privacy-lesson-of-9-11-mass-surveillance-is-not-the-way-forward

Tulley, C. (2013). Migration patterns: A status report on the transition from paper to eportfolios and the effect on multimodal composition initiatives. *Computers and Composition*, *30*, 101–114. https://doi.org/10.1016/j.compcom.2013.03.002

Turner, F. (2015, June 5). *USA freedom act strives to strike privacy-security balance*. TCA Regional News. https://search-proquest-com.eur.idm.oclc.org/docview/1686025649?accountid=13598

Turner, S. (2005). Critical junctures in genetic medicine: The transformation of DNA lab science to commercial pharmacogenomics. *Journal of Business and Technical Communication*, *19*(3), 328–352. https://doi.org/10.1177/1050651905275619

Turow, J., & Draper, N. (2012). Advertising's new surveillance ecosystem. In K. Ball, K. D. Haggerty, and D. Lyon (Eds.), *Routledge handbook of surveillance studies* (pp. 133–140). Routledge.

U.S. Department of Justice. (2015, July 16). *Overview of the Privacy Act of 1974*. https://www.justice.gov/opcl/individuals-right-access

U.S. Department of Justice. (n.d.). *Preserving life and liberty*. https://www.justice.gov/archive/ll/archive.htm

U.S. Department of Labor, Bureau of Labor Statistics. (2018a, Apr. 13). *Accountants and auditors: What accountants and auditors do*. https://www.bls.gov/ooh/business-and-financial/accountants-and-auditors.htm#tab-2

U.S. Department of Labor, Bureau of Labor Statistics. (2018b, Apr. 13). *Network and computer systems administrators: How to become a network and computer systems administrator*. https://www.bls.gov/ooh/computer-and-information-technology/network-and-computer-systems-administrators.htm#tab-4

U.S. Department of Labor, Bureau of Labor Statistics. (2018c, Apr. 13). *Network and computer systems administrators: Summary*. https://www.bls.gov/ooh/computer-and-information-technology/network-and-computer-systems-administrators.htm

U.S. Department of Labor, Bureau of Labor Statistics. (2018d, Apr. 13). *Technical writers: Summary*. https://www.bls.gov/ooh/media-and-communication/technical-writers.htm#tab-1

U.S. Department of Labor, Bureau of Labor Statistics. (2018e, Apr. 13). *Network and computer systems administrators: What network and computer systems administrators do*. https://www.bls.gov/ooh/computer-and-information-technology/network-and-computer-systems-administrators.htm#tab-2

U.S. Immigration and Customs Enforcement. (2020, July 6). *Broadcast message: COVID-19 and fall 2020*. https://web.archive.org/web/20200706195011/https://www.ice.gov/doclib/sevis/pdf/bcm2007-01.pdf

U.S. Immigration and Customs Enforcement. (2020, July 16). *SEVP modifies temporary exemptions for nonimmigrant students taking online courses during fall 2020 semester*. https://www.ice.gov/news/releases/sevp-modifies-temporary-exemptions-nonimmigrant-students-taking-online-courses-during

Van Aubel, P., & Poll, E. (2019). Smart metering in the Netherlands: What, how, and why. *International Journal of Electrical Power & Energy Systems*, *109*, 719–725. https://doi.org/10.1016/j.ijepes.2019.01.001

Vee, A. (2017). *Coding literacy: How computer programming is changing writing*. MIT Press.

Velasquez, M. (2012). *Business ethics: Concepts and cases* (7th ed.). Pearson.

Verhulsdonck, G., Melton, J., & Shah, V. (2019). Disconnecting to connect: Developing postconnectivist tactics for mobile and networked technical communication. *Technical Communication Quarterly*, *28*(2), 152–164. https://doi.org/10.1080/10572252.2019.1588377

Vie, S. (2008). Digital divide 2.0: 'Generation M' and online social networking sites in the composition classroom. *Computers and Composition*, *25*, 9–23. https://doi.org/10.1016/j.compcom.2007.09.004

Vie, S., & deWinter, J. (2016). How are we tracked once we press play?: Surveillance and video games." *Kairos: A Journal of Rhetoric, Technology, and Pedagogy*, *20*(2). http://kairos.technorhetoric.net/20.2/topoi/beck-et-al/vie_dewin.html#main.

Von Drehle, D. (2013, Aug. 1). *The surveillance society: Secrets are so 20th century now that we have the ability to collect and store billions of pieces of data forever*. TIME.com. https://nation.time.com/2013/08/01/the-surveillance-society

Walls, D. M. (2015). Access(ing) the coordination of writing networks. *Computers and Composition*, *38*, 68–78. https://doi.org/10.1016/j.compcom.2015.09.004

Walls, D. M. (2017). The professional work of "unprofessional" tweets: Microblogging career situations in African American hush harbors. *Journal of Business and Technical Communication*, *31*(4), 391–416. https://doi.org/10.1177/1050651917713195

Walsh, L. (2010). Constructive interference: Wikis and service learning in the technical communication classroom. *Technical Communication Quarterly*, *19*(2), 184–211. https://doi.org/10.1080/10572250903559381

Walton, R., Moore, K., & Jones, N. (2019). *Technical communication after the social justice turn*. Routledge.

Walwema, J., Colton, J. S., & Holmes, S. (2022). Introduction to special issue on 21st-century ethics in technical communication: Ethics and the social justice movement in technical and professional communication. *Journal of Business and Technical Communication*, *36*(3), 257–269. https://doi.org/10.1177/10506519221087694

Wang, X., & Gu, B. (2022). Ethical Dimensions of App Designs: A Case Study of Photo- and Video-Editing Apps. *Journal of Business and Technical Communication*, *36*(3), 355–400. https://doi.org/10.1177/10506519221087973

Washington Post. (2013, July 10). *NSA slides explain the PRISM data-collection program*. https://www.washingtonpost.com/wp-srv/special/politics/prism-collection-documents/

Webster, W. (2012). Public administration as surveillance. In D. Lyon, K. D. Haggerty, & K. Ball (Eds.), *Routledge handbook of surveillance studies* (pp. 313–320). Routledge.

Weinberg, J. (2017, July 18). The real costs of cheap surveillance. *The Conversation*. https://theconversation.com/the-real-costs-of-cheap-surveillance-67763

Weller, T. (2012). The information state. In D. Lyon, K. D. Haggerty, & K. Ball (Eds.), *Routledge handbook of surveillance studies* (pp. 57–63). Routledge.

Wells, G., Horwitz, J., & Seetharaman, D. (2021). Facebook knows Instagram is toxic for teen girls, company documents show. *Wall Street Journal*. https://www.wsj.com/articles/facebook-knows-instagram-is-toxic-for-teen-girls-company-documents-show-11631620739

Whalen, D. (2018). Selections from the ABC 2017 annual conference, Dublin, Ireland: Finding a pedagogical pot o' gold. *Business and Professional Communication Quarterly*, *81*(2), 244–265. https://doi.org/10.1177/2329490618766637

Whitehouse of George W. Bush. (2006, Mar. 9). *USA PATRIOT Act*. https://georgewbush-whitehouse.archives.gov/infocus/patriotact/

Wickman. (2014). Wicked problems in technical communication. *Journal of Technical Writing and Communication*, *44*(1), 23–42. https://doi.org/10.2190/TW.44.1.c

Wilson, G., & Wolford, R. (2017). The technical communicator as (post-postmodern) discourse worker. *Journal of Business and Technical Communication*, *31*(1), 3–29. https://doi.org/10.1177/1050651916667531

Wise, J. (2017). Assemblage. In L. Ouellette & J. Gray (Eds.), *Keywords for media studies*. NYU Press. https://doi.org/10.2307/j.ctt1gk08zz.8

Wise, J. M. (2002). Mapping the culture of control: Seeing through *The Truman Show*. *Television New Media*, (3)1, 29–47. http://doi.org/10.1177/152747640200300103

Wise, J. M. (2016). *Surveillance and film*. Bloomsbury.

Woods, Charles. (Apr. 2021). Teaching privacy ethics and identity using digital genealogy databases. *Conference on College Composition & Communication*. Virtual.

Wright, D., Malone, E., Saraf, G., Long, T., Egodapitiya, I., & Roberson, E. (2011). A history of the future: Prognostication in technical communication: An annotated bibliography. *Technical Communication Quarterly*, *20*(4), 443–480.

Wysocki, A. F. (2013). What do technical communicators need to know about new media? In J. Johnson-Eilola, & S. A. Selber (Eds.), *Solving problems in technical communication* (pp. 428–453). University of Chicago Press.

Young, I. M. (1990). *Justice and the politics of difference*. Princeton University Press.

Young, S. (2020). Your digital alter ego: The superhero/villain you (never) wanted transcending space and time? *Computers and Composition*, *55*, 1–11. https://doi.org/10.1016/j.compcom.2020.102543

Young, S., & Pridmore, J. (2021, May 21). Tactical communication and resisting data surveillance. Paper presentation at Data Justice Conference of the Data Justice Lab, May 20–21, Cardiff, UK. (Online). https://datajusticelab.org/live-conference-programme/

Zachry, M. (2008). An interview with Susan Leigh Star. *Technical Communication Quarterly*, *17*(4), 435–454. https://doi.org/10.1080/10572250802329563

Zhang, S., Gosselt, J., & de Jong, M. (2020). How large information technology companies use Twitter: Arrangement of corporate accounts and characteristics of Tweets. *Journal of Business and Technical Communication, 34*(4), 364–392. https://doi.org/10.1177/1050651920932191

Zhang, Y., & Saari Kitalong, K. (2015). Influences on creativity in technical communication: invention, motivation, and constraints. *Technical Communication Quarterly, 24*(3), 199–216. https://doi.org/10.1080/10572252.2015.1043028

Zuboff, S. (2019, June 6). *The surveillance threat is not what Orwell imagined. Time.* https://time.com/5602363/george-orwell-1984-anniversary-surveillance-capitalism

Zwagerman, S. (2008). The scarlet *p*: plagiarism, panopticism, and the rhetoric of academic integrity. *College Composition and Communication, 59*(4), 676–710. https://www.jstor.org/stable/20457030

Index

absence: of surveillance, 29, 111, 114; at work, 76

abuse: of bodies, 83; of power, 21, 32, 68, 110, 123

acceptable: behavior, 84–85, 94, 7; risk, 74; surveillance, 23, 25–26, 35, 39, 97, 103, 115, 135, 162, 165n1

access: to information, 1, 14, 48, 62, 69, 79, 85, 101, 131, 141, 166n1; to technology, 6, 15, 35, 43–44, 120, 145

accessibility, 15, 31, 62, 131, 168n2

accountability, 33, 76, 103, 105, 109, 112, 124

accountant, 46–47, 55–56

act: as a behavior, 22, 34–35, 52, 80, 86, 93, 97, 104, 111–12, 122, 160; ethical and just, 37, 86–87, 89, 115, 126, 160, 164; government, 105–8, 110–13, 123–24, 143–44, 151; of surveillance, 33, 72–74, 76–78, 80, 126, 151, 166n4. *See also* surveillance scenario

Actor-Network-Theory, 135, 169n2

action: ethical, 86–88; socially just, 88–90

adapt: behavior 122, 135; information, 59, 63, 80, 87, 146

advocate: as an action, 31, 37–38, 89, 118, 121, 130, 133, 161, 166n9; as a title, 148, 150

aesthetic: surveillance, 26, 31–32

affect: and effective surveillance, 34–36; to make a difference, 30–31, 46, 51, 94, 112, 134, 158

affordance; 28–29, 122, 124–25, 151; technological, 10, 13, 63, 77

age: as a demographic, 11, 29, 34, 145, 168n6

agent: as employee, 8, 26, 45, 47, 71, 130; machine as an, 135; as site of resistance, 38–39, 117; of surveillance, 12, 30, 33, 36, 38–39, 72–74, 76–78, 80, 117, 125, 127, 131, 134, 136, 139–41, 147, 151, 166n4. *See also* surveillance scenario

agency: feeling of control, 1, 3, 29, 38–39, 72, 117, 131, 137, 163–64; organization type, 1, 14, 42, 44, 51–52, 67, 93, 104, 107, 123–24

Air Force Magazine, 137

Alexander, Keith B., 105

algorithm, 2, 11, 34, 66, 80–81, 83, 103, 135, 140–43, 153, 156–57

Amazon, 12

America, United States of, 5, 22, 28, 44, 86, 93–94, 101, 103, 106–7, 109–11, 123, 137
American Civil Liberties Union, 22, 137
Amnesty International, 123
Andrejevic, Mark, 22, 71, 165n3
anonymous, 3, 12, 25
Anonymous, 120, 128
antenarrative, 30
anxiety, from surveillance, 34, 36
apocalyptic visions of a surveillance future, 161–62
appeal, rhetorical, 24–25, 63
appropriate: suitable, 23, 39, 63, 70, 80, 85, 120, 125, 147; use for one's own, 30, 79, 119, 122
argument: of data protection and privacy, 33, 97; ethical and just, 96, 103, 106, 112; rhetorical, 15, 24–26, 34, 36, 45, 62–63, 77, 136, 146, 163; surveillance as an, 24–27, 34, 45. *Also see* nothing to hide argument
archive, 26
Aristotle, 24
art: surveillance as, 15, 44; tactics as 119, 122
assemblage: collection, 14, 23, 25, 72, 166n4, 167n4. *See also* surveillant assemblage
attitude , 35–36, 76; towards surveillance, 26, 31, 35
audience: algorithms and, 135; argument, rhetoric, and 19, 23–25, 35–36, 48, 52–53, 62–64, 75, 77–79, 87, 144, 154, 157, 160, 166n9; surveillance and the, 27, 34, 78, 143; tactics and, 120, 126, 143
audit, 43, 45, 55, 57, 74, 124
authority, 25, 36, 46, 88, 91, 93, 106–7, 110–11, 129, 134, 142–43, 151, 165n2, 168n3

automated surveillance, 25, 41, 43, 135
awareness: information conscious and, 6, 108, 130, 152, 155, 161–62; socially conscious and, 16, 125, 138, 143, 152, 155, 168n2

Ball, Kirstie, 93
Beck, Estee 18–20, 156
behavior: changes due to chilling effects, 29, 36; ethical, 86–88; oppressive, 90; as a way of acting, 2, 10, 12, 34, 70, 84, 124, 129–30, 143, 145, 163, 168n4
Bemer, Amanda, 47–49, 53, 59
benefit: ethical consideration, 86, 95–98, 103, 105, 111, 114; general advantage, 16, 91–92, 120; of surveillance, 15, 27, 32, 84, 92, 95, 115, 160
Berkeley Center for Law & Technology, 144
bias: algorithm, 80–81, 83; information, 36, 86; viewpoint, 87–88, 157
big brother, 30–32, 46, 147
big data, 58, 80
Bill of Rights, 108
Black Mirror, 32, 162
bodies, 9–10, 17, 19–20, 29–30, 32, 49, 84, 92, 97, 118, 129, 135, 150, 166n7, 166n11, 168n6
body count, 32
Booz, Allen Hamilton, 43, 166n1
border, 13, 17, 19, 34, 37, 85, 93, 111, 138, 156, 162, 165n2
boundary, 15–16, 33, 107, 128
Bowker, Geoffrey C., 6, 138
breastfeed, 17
Brennan Center for Justice, 144
Brock, Kevin, 156
Brumberger, Eva, 75, 79

budget, 45, 102
Bullrun, 65–66. *See also* projects of surveillance
bureaucracy, 45, 67, 72
Burke, Kenneth, 6, 72
Burnett, Gary, 62
Bush, George W., 107, 110
business, 11–12, 24, 42, 45, 56–57, 70, 74, 77–78, 85–87, 118–19, 138, 148–49, 154
business ethics, 85–86

camera, 25–26, 36, 71, 129, 135
cancer, 17
care: elder, 34; ethics of, 84, 95, 98–99, 103, 112–13
Cargile Cook, Kelli, 105, 167n2
categorical suspicion, 162, 169n1
categorize, 11, 16, 45, 54–56, 64, 75, 92, 102, 115, 161–62, 167n5
category, 11–12, 14–17, 19, 26, 33, 56, 95, 97–98, 104, 138
celebrity: surveillance, 140
census, 17, 139
Center on Privacy & Technology at Georgetown, 144
Central Intelligence Agency (CIA), 31, 42, 67, 94, 104, 123
Centre for Research into Information Surveillance & Privacy (CRISP), 137
Chertoff, Michael, 113
chilling effects, 2, 28–29, 35–36, 84, 94, 143, 163
China (Chinese), 17, 44, 64, 122, 166n1
Chokrevski, Tomislav, 29
cinematic: surveillance, 32
citizen, 29, 32, 35, 57, 93–94, 106–7, 112, 123, 128, 149, 160
civil liberties, 22, 28–29, 35, 84, 110–13, 137
class, 91, 93, 147, 162, 168n6

classified information, 1–2, 5, 100–1, 119, 122–23, 141
classify: bodies, 5–6, 42, 84, 92–93, 147; as a process, 19, 24, 44, 49, 55, 74, 83, 91, 135, 141–42, 158, 166n2; as social sorting, 64, 83. *See also* sorting
classroom, 17, 20, 66, 132, 146, 149
clearance. *See* security clearance
closed-circuit television (CCTV), 25–26
coalition, 33, 89, 115, 129–30, 160
code, 53, 66, 135, 138, 140–41, 143, 153–54, 156–58
collect, data and information, 5, 12, 14, 20, 22, 28, 33, 41–42, 46, 55–56, 65–66, 74, 80, 101, 107–9, 135, 138, 145, 165n3
collection, aggregate of things, 25, 30, 165n1
collective, as a group, 18, 48, 65, 86, 89–90, 97, 102, 111, 117, 129, 131
Colton, Jared S., 29, 86, 119–21, 127
Command, Control, Communications, Computers, Intelligence, Surveillance and Reconnaissance (C4ISR), 57
communicate: data and information, 1, 21, 26, 47–48, 50, 52–53, 56, 61, 64, 66, 71, 75, 85, 119 136; using technology, 21, 47, 49, 64, 66–67
community: as a collection of people, 7, 23, 27–28, 30–31, 33, 35, 86, 99, 115, 128
computer, 21, 31, 42–44, 50, 52–53, 58, 64, 66–67, 74, 76–77, 87, 101–2, 104, 126–27, 137, 153, 158
computers and writing, 19, 137
conduct: as a noun describing in one's behavior, 42, 99, 119, 131; as a verb to carry out action, 42, 50, 109, 117, 119–20, 12; privacy,

conduct *(continued)*
  38; surveillance, 5, 27, 30, 36, 38, 46, 57–58, 79–80, 104, 128, 139, 144–45, 158
conform to expectations and norms, 8–9, 118, 162
consensus, 5, 79, 95, 114
consequence: for privacy, 39; of surveillance practices, 2, 19, 27–28, 37, 31–33, 72–74, 76–78, 80, 83–85, 90, 94, 105, 133, 151, 156–57, 162, 166n4. *See also* surveillance scenario
conservation writing, 134, 136, 157–58
Constitution, United States, 28, 106–8, 113, 123–25
constitutional, 106–8, 124
constitutive rhetoric, 7. *See also* rhetoric
construct: identity, 71; ideology, 111; rhetoric, 41, 45; systems, 106; truth, 25, 27; visuals, 26
consumer, 10–12, 34, 64, 68–70, 80, 91–92, 108, 139; surveillance 11, 70, 80, 92
content management, 75, 77
contested concept, 7, 8, 10
context: analysis, 88, 133; and circumstances, 9, 13, 15–17, 27, 29–30, 33, 50, 68, 71–72, 74, 85–89, 111–12, 120, 125, 133, 141, 148, 152, 154, 158, 160, 165n6; rhetorical, 22–24, 62–63; for Snowden's example, 100–4; surveillance needs 22–24, 62–63
contextual, 7, 22–23, 30, 88, 133
contextual integrity, 22–23, 165n6
contradictory: positionality can be, 88; surveillance is, 33, 160, 163
control societies, 10, 12–13
cookbook, 15, 19

corporate, 6, 14, 20, 43, 45–46, 69–70, 87, 130, 139, 157–58
corporation, 5, 10, 12, 14, 23, 26, 30, 36, 78–79, 84, 123, 135, 137–39
counterintelligence, 44
countersurveillance, 37, 125, 140; avoidance moves, 125; blocking moves, 125; discovery moves, 125
course and learning management, 18, 20
course example, surveillance, 146–54
COVID-19, 135, 141, 149, 153, 155, 169n4
cracks, 1, 19, 118–19, 122
creativity, 19, 35, 74, 83, 91, 113, 161, 163
credit, 14, 63, 74
creep, 22, 161
crime, 5, 7, 27, 47, 87, 115, 143
critical, 30–31, 49, 88, 115, 135–38
Cuijpers, Colette, 131
cultural imperialism, 89, 91, 93–94, 112. *See also* five faces of oppression
culture, 10, 29, 30–32, 34, 38, 45–46, 63, 85, 89, 91–94, 112, 121, 128, 168n4, 168n6, 169n
custom, 45
customer, 27, 48, 53, 97, 135, 139
cut-and-paste, 155
cyber, 42, 44, 65, 117

Danielson, Peter, 7, 47
Darmohray, Tina, 43, 52–54
data: agent or collector, 74, 80; analysis and interpretation, 26, 41, 44–45, 56, 58, 62–63, 80, 86, 102, 167n1; anonymous, 12; assemblages, 14; automated, 135; big, 58, 80; bodies, 14, 19–20, 83; collection and gathering, 33–34, 43, 66, 135, 143, 162, 165n3; cut-

and-paste, 155; definition, 167n1; digital, 5, 21, 65, 67, 74, 131; driven decisions, 81; dystopian, 161–62; ethics, 161; exposure, 117, 126; flows, 68; geolocation, 145; health, 138; identifiable, 12, 41, 55, 70, 11; information, 12, 41, 45, 55, 62–63, 67, 75, 102, 163; justice, 150, 161; mining, 92, 157; misuse, 34; permissions, 92; personal, 12, 14, 22, 34, 41, 55, 63, 70, 81, 131, 147, 161; plan, 143; privacy, 22, 38, 81, 97, 117, 131, 144; protection, 33, 81, 97, 117, 143; relation to information, 62–63, 167n1; science, 65; smart, 131, 140; social media, 80; sorting, 63; 167n1; storage, 81, 105; surveillance, 14, 34, 41, 44, 46, 52, 67, 92, 94, 163; synecdoche, 83; TikTok, 145; transmit, 84; utopian, 161

data double, 14, 20, 83

database, 5, 17, 26, 43, 67, 72, 92, 102

datafication, 14, 83

dataveillance, 14, 70, 73–74, 77, 80–81

Data Justice conference, 89

de Certeau, Michel, 19, 118–22, 126, 128, 136, 168n5

de Graaf (2010), 72, 123–24

de Jong, Menno D. T., 79

decade of universal surveillance, 134–35

decentralized, 14

decision-making, ethical and just, 2

Defense Intelligence Agency (DIA), 42

degrading, 29, 32, 92

Deleuze, Gilles, 8, 10, 13–14, 147, 167n4

deliberation, 15

deliverable, 48, 75, 78, 134, 150

Dell, 43

democracy, 33, 35, 103, 109–11, 157

demographic, and categorizing, 11, 26, 29

Department of Homeland Security, 46, 113

design, 8–9, 15, 18–19, 48, 56, 62, 65, 68, 74, 76, 79–80, 85, 96–97, 100–1, 111, 120, 130, 138, 158, 161, 167

detrimental: ethical choices, 88, 104; surveillance as, 23, 29, 32, 103–5, 125, 157, 163

differential: impact, 22, 83; treatment, 93

digital realm, 14, 18–19, 21, 26, 28, 63, 67, 77, 126, 144, 153, 158

Digital Rhetorical Privacy Collective, 18

digitalization, 11

digitized, 22, 159

Ding, Huiling, 121–22, 168n2

disability, 31, 85

disciplinary power, 8–9, 19, 136

disciplinary society, 9–10, 12

discipline: academic 1, 3, 6, 39, 58, 62, 64, 115, 158; Foucauldian 9, 93, 166n11; as control of self or others, 8–12, 19, 32, 39, 88, 127–28, 136

discourse, 6–7, 20, 24, 66, 89, 118, 122, 134, 136, 145–46

discovery moves, 125. *See also* countersurveillance

discursive ethics, 33, 95, 98

Dishfire, 65–66. *See also* projects of surveillance

disproportional: surveillance as, 31, 33, 35, 84, 115, 164; ethical distribution of goods, 98

dissensus, 95, 114

distribute, 11, 13–14, 28, 68, 92, 98, 103, 109, 111, 118, 120

dividuals, 10, 14

docile bodies, 9

document: as a noun, 24, 28, 43, 48, 52–53, 55, 57, 62, 66–69, 75, 101–2, 106, 109, 112, 120, 128–30, 139–42, 145–46, 150, 163, 168n7; as a verb, 93
Dombrowski, Paul M., 85–86
doorbell camera, 26
drone, 26, 28
Dubrofsky, Rachel E., 30–31, 129
Dutch government, 131
dystopia, and surveillance, 32, 161–62

educational surveillance, 68, 127
economy, 11–12, 21, 28–29, 33, 42, 70, 93, 98, 112, 121, 134, 162, 166n10, 168n3
ecosystem, 136, 138, 146, 152
effect: and affective surveillance, 34–37; as consequence, 2, 28, 35, 37, 63, 76, 80, 84–85, 102, 122, 151; and rhetoric, 16, 24, 36, 57, 62–63. *See also* chilling effects
efficacy, 3
efficient, 55, 62, 79, 96
electronic, 19, 44, 66–67, 75, 107, 137, 142, 149, 151
Electronic Frontier Foundation (EFF), 137
email, 65, 69, 75, 107, 139, 142
emotion, 34, 36, 44
employment, 2
empower, 2, 7, 15, 33, 36, 49, 71, 150
enemy, 32; of the state, 5
Enterprise Knowledge System, 5, 100, 102
environment, 15, 18, 64, 71, 79, 134, 157, 159, 162
epistemology, 24–25, 93
equitable, 37, 94, 162
Ericson, Richard V., 14
essay of surveillance. *See* paradigms and essays of surveillance

ethics: action and, 86–90, 115, 127, 161; analysis, assessment, evaluation, and, 23, 33, 69, 73, 81, 87, 102, 114, 131, 133, 161; and business definition, 85; and care, 84, 95, 98–99, 103, 112–13; choices, 87, 138; constitutional, 123; context and situation in, 100; continuous process, 33, 95; decision making and, 2, 33, 88; definition, 95; enticement and, 87; frame and heuristic, 38, 58, 83–86, 95, 104, 114–15, 148, 160; hacker and, 43; 'I was just doing my job' and, 88; imagination and, 165n1; individual, 86; of justice, 85, 95–99, 103, 109–12, 114–15; justifiable, 125; legislation, 151, 155; literacy, 105, 160, 167n2; needed to communicate information, 85; path and, 87; principles, 123; privacy protection and, 145; resistance, 120; rights-based, 33, 95, 98, 103, 106–9, 112; six queries of action, 87–88; Snowden's example analysis and, 100–14, 146; social justice, 6–7, 69, 83, 85–86, 88–90, 114–15, 150; societal, 86; standards, 167n2; and surveillance, 2, 6, 17–18, 23, 26, 32–33, 106, 117, 128, 130, 134, 137–38, 146, 151, 155; tactics and, 127; technical communication and, 2, 32–33; topics of technical communication, 85; use of tech and, 7, 15; utilitarian, 95–97, 99, 103–5
ethnicity, 29, 31, 168n6
ethos, 23, 63, 78, 91, 126, 129
Europe, 45, 86, 143
evaluate: surveillance in context, 23, 30, 38; using ethics to, 32–34, 96, 100, 103, 131, 161; examine

something further, 1–2, 26, 57, 79, 83, 85, 87, 104, 115, 117, 127, 130, 133, 135, 148, 151, 153–54, 160–61, 163, 167n1; one's role in surveillance, 39, 59, 72, 81, 87, 114, 160; performance, 70, 76; technology, 43, 152
everyday surveillance, 6, 8, 12–13, 21, 27, 32, 39, 41, 46–47, 55, 58, 64, 69, 80–81, 94, 131, 135–36, 147, 151, 155, 159, 163
Executive Order 12333, 42
Executive Order 10450, 143
expectation: of future, 89, 119–20, 153, 155, 166n3; institutional, 8, 49, 76, 78; of privacy, 5; of those in power, 8; social, 26, 124, 134
expertise, 21–22, 37, 48, 60, 91
exploit, 89, 91–92, 161. *See also* five faces of oppression
ex-partner, 29

Facebook, 11
facial: recognition, 126; scans, 26
Fairview, 65–68. *See also* projects of surveillance
Family Educational Rights and Privacy Act (FERPA), 144
fear, 2, 29, 34–35, 84, 94, 161–62
Federal Bureau of Investigation (FBI), 8, 42, 45–46, 58, 101, 144
feminist, 30–31
fertility, 17
FISA Amendments Reauthorization Act of 2017, 143
file: document, 53, 66; Snowden's, 43, 45, 67, 75
financial, 10, 12, 55–56, 74, 96
Finn, Jonathan, 25–26, 72
fiscal, 14
fit in, surveillance used to see how bodies, 10

Fitbit, 22
Five Eyes, 5, 101. *See also* projects of surveillance
five faces of oppression, 91, 112
Floyd, George, 154
footage, 25–27, 32
Ford Pinto, 96, 108
foreign, 42, 101, 109, 119, 122–23, 143
Foreign Intelligence Surveillance Act, 143
forum, 19, 120, 125–27, 142, 168
Foucault, 8–10, 12–13, 19–20, 93
Four Rs of Walton et al., 89, 129. *See also* recognize, reveal, reject, and replace
free, 12, 34–35, 97–98, 104, 108, 111–12
freedom, 35, 43, 84, 97, 108–9, 111–13, 169n5
FREEDOM Act, 107–8, 110
Freedom of Information Act, 143
Fuchs, Christian, 12, 101
fundamental, 28, 33, 54, 94, 98–99, 114, 129, 136–37, 147, 159
future, 2, 6, 12, 18, 21, 34, 44–45, 76, 79, 81, 87, 100, 105–6, 132, 138, 144, 146, 155, 157, 161
futures studies, 161–62

Galič, Masa, 10, 13–14, 29
Gallagher, John R., 135
games, 15, 18, 66, 120
Garden & Gun, 34
Gates, Kelly, 22
gather: data, information, and knowledge for surveillance, 10–11, 14, 21–22, 32, 34, 39, 41, 43, 45–48, 56–57, 66–67, 73–74, 81, 100, 102, 106, 108, 131, 135, 139–41, 143, 145, 161–63
gaze: and power, 8, 25, 72; unwanted, 33, 98; at women's bodies, 17

gender, 11, 31, 93, 162, 168n6
genealogy, 17
General Data Protection Regulation (GDPR), 143
genre, 118, 134, 136–39, 141–42, 146, 157, 160, 166n3, 169n3
Gilliom, John, 92–93, 128, 168
Godhe, Michael, 161
Gonzalez, Alberto, 111–12
Goode, Luke, 161
goods, 11, 28, 91, 98, 111, 166n9
Google, 11–12, 35, 126, 157
government: agents of surveillance go beyond the, 13; authority to gather information questionable, 5, 106; authority and power, 8, 10, 36, 156; and care, 113; causes harms to privacy, internet freedom and basic liberties, 109, 113; and civil liberties, 28, 112; culture, 87; and democratic control, 109–10; and Dutch smart meters, 131; enemies aren't the only targets of, 5; hacking, 43; imaginary, 46; information demand, 11; and intellectual freedom, 35; and internet monitoring, 20, 35; officials, 110; has no right to surveil, 108–9; preyed on ignorance, 6, 130; propaganda, 162; property, 121; reporting, 122; right to surveil legally, 106–8, 130; right to surveil morally, 106–8; routine practice of surveillance for, 10; safety or security, 25, 110; speak out against the, 106; surveillance, 13–14, 17, 28–29, 45, 58, 67, 102, 105, 109, 123–25, 128, 130, 135, 137, 141, 149; and total control, 110; transparency, 109, 143; $200,000 value of human life, 96; UK cooperation, 101; watches

students, 156; work, 104, 143, writing, 139
greater good, 96–97, 99. *See also* utilitarian ethics
Greenwald, Glenn, 106
The Guardian, 5, 104, 113
Guatarri, Félix, 14

Haaksma, Tim R., 79
hacker, 26, 43, 168n3
Haggerty, Kevin D., 9, 14, 93, 147
harassment, 19
harm: algorithmic, 2; to the body, 29, 92, 97, 166n7; ethics and, 87–88, 96–97, 115, 148, 160; oppressive, 92, 117; to privacy, 2; rhetoric of 84; of surveillance 27–29, 33–34, 36, 94–97, 104, 106, 117, 127, 133, 150, 160, 163, 165n7. *See also* chilling effects; civil liberties; human rights; social justice
Hawaii, 43, 52
Hayden, Michael V., 109–10
health, 17, 22, 33–34, 36, 65, 92, 96, 98, 138, 141–42
Health Insurance Portability and Accountability Act (HIPAA), 138
helpless, 32, 91
Henning, Teresa, 47–49, 53, 59
heuristic, 2, 84–85, 89, 95, 106, 113, 115, 141
hide, 36, 38, 83, 119, 148
hierarchy, 78, 81
history, 24, 26–28, 33, 63, 74, 88, 93, 101, 123, 146, 167n2
Holladay, Drew, 120
Hollywood, 7–8, 31
Holmes, Steve, 29, 86, 119–21, 127
Holwell, Sue, 167n1
hospital, 9, 13, 30
human, 12, 16, 28–29, 36, 41–42, 45, 48, 70, 91, 96–97, 100, 103, 106–7,

121, 139–41, 152, 156–57, 160, 169n5
human rights, 28–29, 97, 100, 107
human-computer interaction (HCI), 36
hygiene, 39, 117, 131

"I was only doing my job," 88
identifiable: data or information, 12, 41, 55, 70, 100; surveillance, 75; technical communicators, 50
identify: behavior, 10, 84; categories 13, 20, 146, 153; company needs 43; through ethics 95; information, 63, 69, 102; surveillance, 27, 47, 58, 61–62, 64–65, 69, 72, 115, 117, 125, 131, 136–38, 141–42, 148–49, 151, 154, 159–60, 167n5
identity, 19, 48, 130
ideology, 24, 26, 78, 111, 114, 124
illustrative example, 3, 7, 106, 108, 133, 146
imagination, 8–9, 29, 31–32, 39, 46, 55, 83, 161
imbalance, 83, 160
immigration, 17, 95, 153, 155–56, 169
Immigration and Customs Enforcement (ICE), 153–56, 169n4
impact, 2, 6, 15, 22, 25, 31, 35–38, 55, 85, 104, 145, 153–54, 162, 164
individual: as one who is singled-out, 11, 14, 167n7. *See also* dividual
information: architecture, 48, 66, 75, 78, 166n2; communication, 22, 58, 66; contrasted to data 61n1, 63; information communication technology (ICT), 22, 58, 67, 158; definition, 61–62; economy, 11–12, 21; rhetoric, 62–63; surveillance, 63–64; workers, 11, 21. *See also* surveillance scenario
infrastructure, 10, 13, 43–44, 52, 67, 121

infrastructure analyst, 44
injustice: and harm, 84; institutional, 29; and oppression, 89–90; social 86, 94, 150
Instagram, 94
institution: de Certeau and the, 118–21; as a location, 8–10, 12–13, 16–17, 29–30, 32, 34, 36, 46, 68, 71, 76, 78, 80, 85, 87, 89–90, 93, 118–22, 124–26, 128–29, 136, 151–52, 156, 168n1; power and control, 13, 36, 46, 89, 118, 122, 128, 136, 152
instruction: definition of, 68; to direct action, 47, 49, 52–53, 57, 68–69, 75, 156; manuals, 120, 125; of surveillance practices, 68–69
intelligence: act, 143; agency 42, 51, 123; analysis, 42; committee, 109; community, 1, 42, 44, 46, 58, 71, 81, 123, 125; information, 44, 70, 93, 101, 109, 135, 138–39; officials, 42, 107, 110; processes, 109; programs, 57; systems, 57, 142
intercultural, 16, 31, 111
interdisciplinary, 3
interface, 18
internalized, 9
international, 31, 86, 122, 125, 151, 156
internet, 20, 35, 44, 65, 67–68, 70, 72, 84, 100–1, 107, 109, 113, 120, 123, 126, 142, 168n2
internet of things, 84
interpret, 6, 23–24, 30, 41–42, 44–45, 49, 62–63, 68, 74, 87, 123, 130, 154, 156, 167n1
investigation, 1, 29, 70, 95, 139–40, 143, 163
investigator, 1, 75, 111
invisible: drone presence, 94; perspectives of oppressed groups, 91; systems, 2

invisible digital identity, 19. *See also* data double
Iran, 44

Jaeger, Paul T., 62
job, 16, 21–22, 31, 36, 42, 46–47, 49–58, 62, 69, 76, 88, 94, 98, 104, 112–13, 127, 130, 138, 141–42, 144–45, 149–50, 152, 154, 166n1, 166n2
Johnson-Sheehan, Richard, and Morgan, Larry, 134–36, 146
Jones, Natasha N., 29, 31, 37, 84, 86, 88–89, 91, 129, 148, 150, 155
journalism, 1, 67, 69, 75, 106, 108, 113, 121–23, 129–30, 138, 140–41
just, morally, 2, 23, 37–38, 86, 88–89, 94, 109, 111, 114–15, 152, 161–62
justice: American 111; data, 150; definitions of 98, 109; democracy and, 109; environmental, 134; ethics and, 85, 95–99, 103, 109–12, 114–15; and George Floyd, 154; intercultural looks at, 111; and less surveillance, 111; and oppression, 112; questions of what is, 37, 112, 137, 168n2; and the state, 9; through sustained surveillance, 110–11; system, 155, 169n3; three types of justice, 98–99. *See also* social justice

Karreman, Joyce, 79
Kimball, Miles, 16, 47, 119–21, 128, 166n3
knowledge: awareness and skill, 43–45, 48–49, 53; create, 21, 24; communicate, 21, 54, 120–22, 125–26; 81, 91–92, 96–97, 100, 102–3, 112, 118, 120 126–27, 143–44, 149, 157, 163; 167n2; gather, 39; management, 64; program. *See also* Enterprise Knowledge System

Koops, Bert-Japp, 10, 13–14, 29, 131
Kreibich, Rolf, 161

la perruque, 126–27
labor, 2, 43–45, 58, 92
lateral surveillance, 70–71, 78–79
Lauer, Claire, 75, 79
law, 28, 45, 57, 74, 86, 98, 103, 106–7, 111–12, 122, 125, 136, 143–44, 146, 148, 151, 156
law enforcement, 22, 26, 71, 107, 110, 135
learning and course management, 18, 20
legal: by law, 31, 97, 102, 106–7, 112, 122–24, 134, 137, 145–46, 152–53, 155–56; right, 29, 93, 97, 106–8
legislation, 18, 38, 105–8, 110, 139, 143–44, 146, 151
legitimate, 5, 25–26, 48–49, 84, 91, 93–94, 118, 122, 144, 156–57
lens: cultural, 30, ethics as a, 97; oppression as a 11; panoptic, 13, 20; social justice, 29, 88–89; surveillance as a, 2, 6, 25, 31, 133; terms as a 6; technical communication as a, 59, 62
letter, 75, 126, 139
Library of Congress, 66
limits, 5
Lindgren, Chris Aaron, 156
linear, 14
little brother, 30
Lizzo, 129
loyalty card, 10
Lucas, George R., Jr., 102
Lyon, David, 12–13, 41, 55, 84, 92–93, 147, 165n1, 166n11, 169n1

MacDonald-Evoy, Jerod, 126
machine, 17, 26, 109, 113, 135, 160, 163
machine learning, 11, 140

Magnet, Shoshana Amielle, 30–31, 129
Malone, Edward A., 120
manage: bodies and others, 21, 30, 52, 54–57, 81; content, 75, 77, 140; data, information, and knowledge, 16, 21, 30, 57–58, 64, 77, 134; legally, 153; organizations, 55; risk, 64, 105; social media disclosure, 35; strategies, 118, 120; technology, 16, 43, 54
manager, 21, 54–55, 128, 135–36
manipulate, 21, 81, 118, 121, 155
manuals: technical, 42, 75
marginalize, 27–28, 89–92, 119, 122, 161–62, 166n9. *See also* five faces of oppression
Marine Corps Intelligence School, 57
Markel, Michael, 122
market: futures, 12; price, 11; product, 34
marketing, 22, 67, 71
Marx, Gary T., 7, 33, 69–70, 74, 125, 148, 167n3, 167n6
mass surveillance, 1, 5, 28, 89, 100, 103–6, 108–14, 123, 125, 130, 135
matrix of technical communicators and surveillance workers, 55, 56–56
McQuade, Samuel C., 7, 47
media, 2, 5, 10, 12, 15, 30, 35–36, 38, 45–46, 48, 53, 58, 66, 70, 72, 80, 92, 94, 101, 110, 121–22, 125, 137–41, 143, 145, 153, 158, 160, 166n8, 168n2. *See also* social media
medical: care, 98; device, 135, 142; doctor, 59; information, 141; procedure, 64; records, 141; surveillance, 17, 135, 141; writing, 142
memo, 75, 139, 150, 156
mental, 21, 36, 94, 98
message, 16, 24, 63, 79, 87, 145, 153, 155, 168n2

metadata, 5, 22, 80, 100, 102, 140
meticulous ritual of power, 39
Miceli, Marcia, P., 122, 124
microtechniques of control, 39, 136, 166n11. *See also* meticulous rituals of power
mobility, 45–46, 56
Monahan, Torin, 128, 168
monitoring: of accounts, 45; of borders, 13, 93; of citizens, 93; constant, 10, 93; and control, 10, 139, 140, 142, 163; of customers, 139; dataveillance and; 70; devices, 13, 22; espionage and, 70; everyday, 169n1; fear and, 84; focused or narrow, 28, 47; government, 8, 146; grades and, 136; health, 22, 141; institutional, 8; internet, 20, 70; keystroke, 71; legitimize, 26; panoptic, 20; person who does, 17, 135; personal data, 70; power and, 93; productivity and performance, 92, 127, 139; social media, 70, 101, 139; surveillance practices and, 161; surveillance workers and the art of, 44; surveillance writing documents, 134, 142; suspects, 136; targeted, 7; technologies of, 127, 140, 149; too much, 20; unknown, 20; user experience and, 58; utility and, 34; visible or subtle, 163; workplace, 70, 92, 127, 139; writing center, 20
Moore, Kristen R., 31, 37, 84, 86, 88–89, 91, 129, 155
motivation: ethical, 113; in scenarios of surveillance, 33–34, 72–74, 76–78, 80, 151, 166n4; surveillance capitalism, 34; tactical, institutional, and strategic, 120. *See also* surveillance scenario
movie, 31, 46, 166n7
mundane: surveillance is, 13, 163; tasks, 43

MYSTIC DCSN, 65–66. *See also* projects of surveillance

narrative, 14, 24–25, 41, 63, 117, 149, 166n3. *See also* antenarrative
national security, 5, 84, 94, 103, 112, 123, 167n1, 168n8
National Geospatial-Intelligence Agency (NGA), 42
National Reconnaissance Office (NRO), 42
National Security Agency (NSA), 1, 5, 42–44, 51, 65, 67–69, 101–5, 109, 118, 122–24, 128, 146
Near, Janet P., 122, 124
negative: rights, 97; surveillance as, 32, 84, 148, 160, 163
network, 13, 19, 34, 42–44, 67, 71, 102, 104, 137, 145, 152–53, 158
neutral, 12, 32
New York Times, 67, 102
New York University Law, 144
Newell, Bryce Clayton, 29
Nineteen Eighty-Four (1984), 31–32, 46, 162
normalized: oppression, 155; surveillance, 26–27
norms, 9, 23, 26, 32, 45, 52, 91, 97, 161
nothing to hide argument, 83, 148

Obama, Barack, 105
obfuscation, 15, 137, 144
observe: the body, 166n11; culture, 128; in the field, 56; in reality, 6; rhetoric, 24; as surveillance, 45, 142; suspects, 7; workplace, 70
observer: as agent of surveillance, 74. *See also* surveillance scenario
obtain: communication, 100; information, 33, 42, 44, 46, 67, 81, 134, 142; skills, 53

Oertel, Britta, 161
online, 2, 5, 15, 20, 29, 36, 42, 77–80, 90, 94, 101–2, 121, 125–27, 142, 153, 155–56, 169n4
online games, 15
oppress: and ethics 95, 114; as harm, 84, 94, 117; recognize, reveal, reject, and replace, 89–90; and Snowden 104; and social justice, 37, 84, 88, 91–92, 112, 114–15, 117, 129, 135 155, 166n9; with surveillance, 3, 11, 32, 37, 84–85, 88, 90–92, 112, 114–15, 121, 129, 135, 150, 155; systemically, 29, 89, 128, 150, 163; utopian absence of, 161; Young's five faces, 91–92, 112. *See also* exploit, marginalize, powerless, cultural imperialism, and violence
organization: place, 24, 42–44, 54–56, 70, 72, 74–75, 77, 85, 91–92, 112, 118–20, 122, 128–30, 137, 149
organize: to order, 2, 53, 78, 81, 92, 118, 121, 150
Orwell, George, 31, 137

pandemic, 138, 141, 156
panoptic, 9, 13, 19–20, 39, 93, 165n2
panopticism, 8–10, 19, 137
panopticon, 8–9, 13–14, 19–20, 32, 46, 147, 165
paradigm: of privacy, 131; and essays of surveillance, 8–12, 14, 32–33, 70, 72–74, 76–78, 80, 151, 166n4. *See also* celebrity surveillance; control societies; consumer surveillance; data surveillance; dataveillance; everyday surveillance; government surveillance; lateral surveillance; panopticism; participatory surveillance; peer-to-peer, self-surveillance; sports surveillance; sousveillance; state surveillance;

Index | 213

surveillance capitalism; surveillance scenario; workplace surveillance
Parker, Donn B., 87
participant, 7, 25, 81, 121, 129, 160
participate, 12–13, 15, 33, 45, 69–70, 75, 89–90, 104, 123–27, 137, 147, 149
participatory surveillance, 70–71, 77, 80, 137
Patriot Act, 105–7, 110–13
pedagogy: surveillance, 18, 64, 136, 146; technical communication, 64, 136, 146
peer-to-peer surveillance, 14. See also lateral surveillance
performance review, 54, 70, 136, 139, 141, 149
period tracker, 17
perspective, 14, 85, 91, 97, 99–100, 103–4, 111, 121, 126, 142–43
persuade, algorithms, 153; and document, 75, 154; and ethics, 87; and rhetoric, 24, 63, 158
Pflugfelder, Ehren Helmut, 72, 120
pharmaceuticals, 17
Phoenix police, 155–56
phone: as communication and media, 168n2; instructions for the, 68; and privacy, 35, 145; as surveillance site, 5, 22, 44, 65, 67, 100, 102, 107, 119, 126, 145; and tactics, 119, 128
physical, 10, 13–14, 26, 29, 31, 34–35, 42, 61, 75, 92, 94, 118, 158–59
Pinwale, 65–66. See also projects of surveillance
plagiarism, 20, 66
plan, 56, 130–31, 143, 150, 160–61
plantation, 19
platforms, 15, 35, 53, 67, 92, 102, 121, 126, 145, 158
police, 7, 23, 27, 47, 135, 139, 143, 154–55, 169n3

policy: documents, 57, 139, 141; ethics and, 96; privacy, 145, 153, 81; as a rule, 42, 45, 58, 65, 104, 139, 156, 169n4; surveillance, 17, 59, 133; workplace and organization, 19, 129
political, 29, 93, 95, 101, 111–12, 121, 134, 137, 140, 146, 152–53, 168n3, 168n8
Popular Mechanics, 137
Popular Science, 137
population, 28, 30, 32, 46, 93–94, 98, 123, 129, 138
positionality, 23, 28, 88, 160
positive: dismissed arrests as, 155; false, 83; interactions, 52; liberty enjoyment is, 112; national security as, 84; surveillance consequences as, 32, 36, 80, 148, 160, 163; type of rights, 97
postmarket surveillance, 17, 135, 141–42
postmodern, 13, 85
post-panoptic, 13
power: abuse of, 110; army and, 118; balance of, 79, 83, 88, 105, 109, 121, 128, 160; of businesses, 118; city and, 118; collective, 129; course/learning management and, 20; critique of, 89; cultural, 128; de Certeau and, 118–22; dimensions of, 89; disciplinary, 8–9, 12, 19; can equalize surveillance, 34; ethics and, 88; everyday, 155; flows of, 88; foreign, 119, 122; Foucault and, 8–9, 12, 19, 93; from below, 71; games and, 18; to gaze, 8, 72; government and state, 93, 109–10, 112, 155, 159, 169n3; helpless without, 91; imagination and, 31; institution and 46, 118, 122, 136, 152; legal, 110, 113; legitimacy

power *(continued)*
and, 49, 118; liberty and, 112; meticulous rituals of, 39, 136; of choice, 112; panoptic, 8–9, 14, 19–20, 93; parental, 162; physical locations of, 159; position, privilege, and, 29, 88, 114;privacy and, 8; proprietary, 118–19; relationships and, 37, 71, 118, 129, 155; rhizomatic, 159; schools and, 156; strategies and, 118–19, 140; structures, 9, 12, 14, 36–37, 166n9; surveillance and, 12, 31, 37, 49, 88, 93, 105, 129, 134, 137, 161; tactics and, 122, 128; those in positions of, 8–9, 39, 45–46, 88, 93, 103, 105, 109, 112, 115, 118–19, 121–22, 140, 160–61; who (doesn't) have, 89, 119, 121; writing and, 169n3. *See also* sousveillance

powerless, 2, 6, 89, 91, 93, 112, 139, 156, 161–62. *See also* five faces of oppression

practice, 5–7, 10, 20–21, 26, 28, 58–59, 90, 105, 118, 127, 134, 142, 146, 148, 150, 153–55, 169

predict: with analytics, 161; behavior, 12; consumption, 34, 64; with policing, 157; with surveillance, 45

press: foreign, 123

printed, 15, 75

PRISM, 5, 100–1. *See also* projects of surveillance

privacy: 2, 5, 8, 18, 20, 22–23, 29, 33–39, 57, 65, 81, 97–98, 100, 104–5, 109, 113, 117, 123, 125–26, 131, 134, 136–41, 143–48, 150, 152–54, 165n6, 166n7, 166n10; definitions of, 38; difference between surveillance and, 20, 37–39, 117, 144–45; harms, 2; paradigm, 131; policy, 145, 153; protection, 38, 144–45, 166n7; and voyeur, 38

Privacy Act of 1974, 143

Privacy International, 137

privacy writing, 144–45

private: experience, 12; information, 128; institution, 32, 71; life 2, 29, 131; property and places, 13, 121

privilege, 29, 88–89, 112, 114, 166n10

problematic: behaviors, 88, 127, 130; to define surveillance without power or too vague, 12, 59; and ICE, 156; surveillance, 7, 28–29, 34, 81, 85, 96, 114–15, 131, 133, 145

problem-solving, 15, 37

process: business and organizational, 54, 57; categories and sorting, 44, 92, 167n5; ethical, 33, 87, 95; grading, 136; and information, 21, 46, 62–63; rhetorical, 63, 77, 167n1; surveillance, 15, 22, 26, 38, 59, 64, 68, 75, 81, 104, 142, 145; technical communication, 48, 85, 156; technological, 41, 79; writing, 62, 155

processing: of information, 22, 57, 75; institutional, 93; of personal data, 12, 41, 55; of surveillance content, 81; word, 148

produce: date, 21; documents, 48, 55, 149–50; societal outcomes, 95–96; tactical opportunities, 122; workers, 49

producer: content, 80, 128; digital 2, 144; and editor, 25; technological, 49

product: for consumption, 11, 34, 79–80, 90, 139, 143; as deliverable, 75, 79–80, 146; information as a, 11, 78; privacy as a, 38; writing as a, 134

productivity, 27, 43, 70, 76, 84, 127

professional: community, 47–48, 53, 72; communication, 31, 136; organization, 43, 164; a person who

is a, 49; as a quality of work, 64, 67, 77, 144, 155, 167n2
profile: employee, 79; high, 2, 25; personal, 79
proliferate: focusing on positive consequences helps see why surveillance does, 32; institutions encourage surveillance to, 32; surveillance does, 28, 32, 115, 155; surveillance enabling technologies, 165n3; of technology, 107; surveillance symbols, 26
projects of surveillance. See Bullrun, Dishfire; Enterprise Knowledge System; Fairview; Five Eyes; MYSTIC DCSN; Tempora; XKeyscore; PRISM
ProPublica, 67–68
protect: data and information, 2, 11, 33, 97, 117, 140, 143, 145, 151; ethically or morally, 108, 145; legally, 97, 143–45, 156; national interests, 42, 110, 168n8; persons, 103–4, 111, 123, 136, 151; privacy, 28, 38, 144; privilege, 89; rights, 28, 97, 107, 111; surveillance, 38
protest, 5, 37, 154–55
public, 2, 5, 31, 33, 44, 67, 69, 71, 77, 93, 106, 109, 110–12, 121–23, 125–26, 130, 135, 137, 142–43, 157, 163, 168n2, 168n7
publication, 2, 53, 66, 120, 125, 137, 169n1
punctuated equilibrium, 37
purchase: as in a job duty, 52; past, present, and future behaviors of, 10–11, 34; shopping transaction, 34, 108; surveillance technology, 71

race: for categoric sorting, 11, 30–31, 91, 93, 162; critical theory, 30; and culture, 31, 168n6; and differential treatment, 93. *See also* demographic

Radio frequency identification (RFID), 18
RAND Corporation, 65
range: of documents, 75; of ethics and rights, 33, 106–8; of intelligence, 42; of rhetoric, 63; of strategies and tactics, 121; of surveillance consequences, 80; of surveillance information, 69, 145; of surveillance symbols, 26; of surveillance work, 45–47, 55, 57, 64, 71, 74, 81, 83, 159; of surveillance workers, 45–47, 56, 61; of a technical communicator, 2, 41, 47, 50–51, 61, 83; of technical communication scholarship, 15, 17, 159; of technical communication work, 41, 50–51, 55, 57, 120, 159; of usability, 79; of what oppression means, 90; of work, 43, 53, 64, 142, 151
reality, 6, 25, 32, 45, 87, 92, 131, 135, 161, 165n1
reality show, 25
Reardon, Daniel C., 120
recognize: context, 23; ethics, 86–87; faces, 126; four Rs, 89, 114, 129; injustice, 29, 89; oppression, 37, 89; Others, 92; positionality, 23; powerlessness, 91; resistance and countersurveillance, 37; rhetoric, 87; surveillance, 1, 7, 27, 37–38, 56, 65, 114, 129, 160–61, 163; surveillance writing, 138; tactical right moments, 122. *See also* Four Rs of Walton et al.
record: data and information, 23, 36, 70, 127; others, 12, 23, 21, 25, 37, 134
records: of behavior, 76–77, 123; financial, 55, 74; fitness device, 22; keeping of, 67; medical, 141; public, 109; of surveillance, 139; telephone, 5, 107

redaction, 143–44
reduce: agency, 39; disempowerment, 119; the impact of surveillance, 2, 46; surveillance, 38, 127, 131; visibility, 117
regime, 31–32, 102, 157
reject: employer strategies, 122; Walton et al.'s Four Rs and surveillance, 36–37, 89–90, 114–15, 129, 131, 133, 160, 162. *See also* Four Rs of Walton et al.
religion, 29, 168n6
replace: surveillance and Walton et al.'s, 4Rs, 37, 90, 115, 129–31, 133, 152, 160, 162; tactics for strategies, 120. *See also* Four Rs of Walton et al.
resist: agents and targets can both, 30–32, 117; arguments, 59, 123; and culture, 128; and ethics, 58, 127, 131; as an individual act, 2, 58, 117, 131; rhetorics of surveillance, 5–6, 25; through social coalitions, 58, 117, 129–31; strategies through tactics, 118–22, 125–27; surveillance, 1–2, 5, 7, 14–15, 35–39, 58–59, 73, 81, 115, 117, 123, 125–27, 131; with tactical communication, 125–28; as a whistleblower, 121–25. *See also* countersurveillance; Four Rs of Walton et al.; tactical communication; whistleblowing
reveal: make visible, 65, 101–2, 112, 114, 128; personal details, 131; surveillance, 76, 118, 133; "truth," 26; Walton et al.'s Four Rs and surveillance, 37, 89–90, 114, 129; and whistleblowers, 124. *See also* Four Rs of Walton et al.
revenue, 80–81, 55, 78
rhetoric: algorithms and code, 153, 156, 158; is action-oriented, 34; and argument, 15, 45, 63; of care, 84, 95, 162; and composition, 158; constitutive, 7; and context, 23–24; and content management, 77; and data interpretation, 41; definition of, 24–25, 62, 167n2; extraorganizational, 72; of happiness, 112; of harm, 84; of justice, 111; of national security, 94; and situation, 16, 24, 33, 87–88; and skills, 49, 55; and sorting, 64, 167n1; and surveillance, 6–7, 24–27, 31, 33–4, 62, 77, 114, 130, 160–61; of terrorism, 93; us versus them, 94; usability and user experience, 79
rhetorician, 23–24, 27
rhizomes, 14
rhizomatic, 10
Richards, Neil M., 35, 148
rights, 6
rights-based ethics, 33, 95, 98, 103, 106–9, 112
risk: analysis, 102, 139; and creativity, 83, 163; the fear of taking a, 20, 83, 163; financial, 12; categories of, 14, 30, 33, 63, 74, 93, 102; management of, 64, 105; mitigation of, 107; surveillance for, 17; utilitarian ethics and, 96, 108
Routledge Handbook of Surveillance Studies, 136–37
routine: discipline through, 39; surveillance maintained through, 10, 12, 93, 134, 163; tasks, 21
rule, 45, 68, 85, 112, 118–19, 121, 127

safety: as a benefit of surveillance, 84, 103, 111; as entitlement, 110; standards, 96; terrorism's risk to, 17
Sarat-St. Peter, Hilary, 68, 120, 122
school, 9, 13, 16–18, 30, 49, 57–58, 155–56

Seawright, Leslie, 169n3
search, 26, 28, 35–36, 69, 78, 84, 101–2, 126, 142, 144
security: clearances, 123, 149, 163; domestic and homeland, 5, 112; foreign threats and, 42; as free from danger, 71, 84, 103, 107, 143, 169n5; governmental, 25; officials and workers, 7, 43, 47; purposes and measures, 43, 64, 139n; as a specialized and technical topic, 65; state, 123; technology and, 43, 104; *See also* Department of Homeland Security; national security; National Security Agency
seeing surveillantly, 7, 37, 61, 72, 133, 160
Seigel, Marika, 120
self-surveillance, 14, 18
sensemaking: categories and, 78; information and data and, 63; and rhetoric, 24; surveillance and, 41, 45; terministic screens and, 6
sensitized, 2
sensitizing concept, 7
severity, 32, 99
sex: demographics, 11, 162, 168n6. *See also* demographics
sexualized, 19, 129
shadow, 8, 26
share: data and information for visibility, 22, 57, 78, 89, 106, 119–21, 123, 126, 140, 154, 161; duties of surveillance and technical communication, 57; goals for systemic change, 28; knowledge with coworkers and partners, 54, 101, 110; for resistance, 131; surveillance narrative, 41; through tactics or unofficial means, 120–21, 125–26, 140, 166n3
site: of activism, 137; of agents, 147; of confinement, 13; network, 43, 73; and scenarios of surveillance, 33, 72–74, 76–78, 80, 151, 166n4; social media, 10, 53, 66, 80, 166n8; of surveillance, 10, 13–15, 30, 80, 137–41, 151; of work, 43, 69, 144. *See also* surveillance scenario
situated: code's relationship to writing, 156; surveillance is contextual and, 22–23, 30, 33; technical communication is, 16, 85
situation: 1, 9, 16, 22–28, 30, 33, 36–37, 57, 59, 63, 73–76, 84, 86–88, 95, 100, 104–5, 109–10, 114, 117, 122, 151, 160–61, 166n4, 168n2. *See also* rhetoric and situation
Skorvanek, Ivan, 29
slavery, 19
smart meter, 131
Smith, Gavin J. D., 44–46, 58, 149
Snowden, Edward, 1–3, 5–7, 20, 28, 31, 36, 41–45, 51–52, 54, 57, 64–69, 81, 89–90, 93–95, 100–6, 108–10, 112–14, 118–19, 121–27, 129–30, 133–35, 141, 146–47, 149, 151–54, 162–63, 166n1, 168n3, 168n7, 168n8; resume, 42–44, 57, 79
social context, 15–16
social impact, 6, 15
social justice: and action, 86, 88–90, 129; algorithms and code, 156–57; as a collective, 97, 129; and data justice, 150, 153, 161; definition, 29; and ethics, 7, 58, 69, 83, 85–86, 97, 99, 114–15, 148, 150, 161; harms, 28–29, 160; as a heuristic for surveillance, 2; and Immigration and Customs Enforcement, 155–56; and justice, 112; oppression and surveillance, 90–95, 112, 115; and protests, 154–55; and resistance of surveillance, 37, 117, 131, 150; rhetoric and, 16, 24, 33, 87–88; and scholarship, 11, 29, 33, 84, 99,

218 | Index

social justice *(continued)*
    117, 135, 137, 152–57, 160, 166n9, 166n10, 168n2; systemic, 129; and tactics, 129, 150; and targets of surveillance, 30, 36. *See also* coalition
social media, 10, 12, 15, 35–36, 38, 45, 53, 58, 66, 70, 72, 80, 92, 94, 101, 137, 139–41, 143, 145, 160, 166n8. *See also* sites and social media
social sorting. *See* sorting
societies of control. *See* control societies
Society for Technical Communication (STC), 47–50, 52, 62, 64, 66, 68, 166n2
software, 20, 43, 54, 56, 66, 71, 76–77, 79, 92, 135, 142, 158
Solove, Daniel J., 148, 151, 165n7
sorting: definitions of, 63, 92; and discrimination, 83; of information, 75, 80–81, 167n1; of people and society, 11–12, 14, 83, 92, 129, 137, 156; and rhetoric, 63–64, 167n1; for risk and reward, 63–64, 92–93; and surveillance, 30, 41, 63, 74, 165n3
sousveillance, 15, 70–71, 78–79, 140
sports surveillance, 140
stakeholder, 10, 23, 50, 56, 86–87, 96, 99, 130, 134, 157, 168n2
standard, 85, 96–98, 118–19, 143, 145, 167n2
Staples, William, 39, 136, 139n
Star, Susan Leigh, 6, 138
state, the, 5, 9–10, 12–14, 17, 28–30, 32–33, 38, 67, 81, 93–94, 98, 101, 115, 123, 137, 147, 149, 151, 155–56, 159–60
stereotype, 11, 46, 55, 87, 91
Stoddart, Eric, 33, 95, 148

storage, 22, 37, 65, 81, 105, 131, 136, 165n3
store: as a collection, 42; as in commercial building, 13, 72, 74, 84, 139n; data, 12, 21, 28, 61–62, 66, 77, 134
strategy: action plan, 20, 54, 84, 133, 164; business and institutional, 45, 118–22, 125, 127, 129; of conservation writing, 136; rhetoric and, 24–25, 62–63; of surveillance writing, 136–46; tactics and de Certeau, 118–22, 125, 127, 129
structure: argument, 15, 62, 68; change, 129; government and, 67; oppression, 29, 89–90, 155; organization, 78, 128, 138, 163, 168n6; pay, 28; power, 9, 12, 14, 36–37, 88, 160, 166n9; surveillance, 101, 127; and theory, 149
supervisor, 76, 124, 138
surveillance: action plan, 160–61; benefits, 33–34, 84; as cultural, 30–32, 128; definitions of, 6–8, 12, 20, 41; effect and affect, 34–36; enabling technology, 18, 22, 67; as fuzzy, 1, 104, 127; as good or bad, 2, 23, 28, 32, 34, 90; as harmful, 27–30, 84, 127, 150; as helpful, 23, 33–34, 90, 103, 115, 148, 163; as oppressive, 92–94; reduce the amount of, 38, 127, 131; rhetoric, 24–27; scholarship and field, 3, 7–8, 12–13, 136, 159, 163; technologies, 18, 66, 137. *See also* paradigms of surveillance; privacy differences; privacy and surveillance differences; projects of surveillance
surveillance capitalism, 8, 11–12, 32, 34, 70, 150, 165n5
surveillance economy, 11

surveillance scenario, 57, 61, 63–64, 69, 72–73, 75–76, 81, 125, 127, 151, 155, 166n4. *See also* agent, act, site, target, motivation, information, paradigm, and consequences
Surveillance Studies: A Reader, 137
The Surveillance Studies Reader, 137
Surveillance Studies: An Overview, 137
Surveillance Studies Network, 137
surveillance types. *See* paradigms of surveillance
surveillance work: definition, 41; participating in, 21, 31, 41–42, 46, 51, 57, 59, 65, 67, 81, 100, 104, 130, 149, 159, 163
surveillance worker: definition, 41–47; definition, 44; difference between Surveillance Worker and surveillance worker, 44–47; as one engaged in surveillance work, 2–3, 6, 11, 13, 37, 49, 51–53, 55–59, 61, 64, 69, 71, 81, 104, 121, 135, 142, 149, 157, 159, 163, 167
surveillance writing: definition, 134–35, 141; pedagogical strategies, 136–46
surveillant assemblage, 14
surveillant imagination, 8, 31, 39, 46, 55, 165n1
surveillant machine, 17
suspect: a belief, 88, 115; one under suspicion, 7, 27–28, 45–46, 57, 103, 124
suspicion: categorical, 162, 169n1; distrusted, 5, 28, 102
sustainability: as a benefit of surveillance, 34; and conservation writing, 158; social, 169n5; and surveillance, 162
sustainable surveillance, 162
synecdoche, data as a, 83

symbol, 21, 25–27, 31, 167n1
system: access the, 43–44; accountant, 46; architecture, 43; classifications, 83; communications, 62, 66; computer, 43–44, 50, 52–54, 101, 104; course/learning management, 18, 20; critical thinking, 49; culture as a, 168n4; data, 14, 70, 135, 162; educational, 108, 127; ethical analysis, 33; evaluation of surveillance 161; files, 45; financial, 10; identification, 93; information, 6, 62, 107; intelligence, 57; justice, 155, 169n3; legislative, 97; metadata, 102; oppressive, 91, 128; optimizing, 54; organizing a, 118; phone, 44, 107; political, 121; programmers, 54; protest, 37; public-facing, 77; radar, 142; record-keeping, 67, 123; resistance to a, 128; searching a, 102; smart meter, 131; social justice to challenge a whole, 129, 160; social media labor, 92; social, 151; strategies, 118; surveillance, 1, 18, 37, 67, 85, 92, 100–1, 103–6, 108–10, 115, 118, 124, 126, 131, 160, 163; tactics, 122, 127; training for, 54; transparent, 2, 130; user experience and, 79; visible, 2, 6, 9, 37, 163; watching, 12, 36; workplace, 119
system administrator, 42–44, 50–54, 57
systemic: data use, 70; exclusion, 94; exploitation, 89; oppression, 29; visibility, 37; resistance, 117, 160

tactic: always attached to strategies, 120; as collective, 117, 129, 160; context needed to evaluate, 125;

tactic *(continued)*
  as cultural reflections, 128; de Certeau's theories of strategies and, 19, 118–22, 128; definition, 119; delivery of, 126; ethics, 127–28; everyday, 126–28; general method to accomplish something, 54, 71, 104, 143–44, 149; as individual, 117, 129, 160; information, 125; and the institution, 36, 119–21, 123, 126–27, 129, 140, 144, 168n1; and the internet, 126; and materials, 136; and power, 37, 71, 118–19, 122, 128–29, 140; privacy, 125, 143; range varies, 120–21; and resistance, 37, 117–18, 122, 125, 140; and social justice, 129–31; through stories, 166n3, 168n3; and surveillance, 121–24, 126–32; and surveillance cracks, 118–19, 122; vampirizes discipline and control, 127; of the weak, 119, 121–22; whistleblowing as a, 118, 122, 124–25, 138; writing, and the self, 140
tactical communication, 36, 71, 117, 120, 126, 128, 131, 140, 150, 168n1, 168n3; definition, 120; and technical communication, 120
target: body is a, 166n1; often on certain communities, 23, 28, 31, 84, 123; consequences of surveillance are a moving, 2; de Certeau's strategies and the, 118; government's, 45; institution as a, 78; is responsible for privacy, 38, 131; if surveillance is appropriate is a moving, 23; and surveillance scenarios, 33, 72–73, 151, 166n4; by surveillance, 28, 30, 33, 36, 38–39, 44, 73–74, 76–78, 80–81, 89, 94, 101–3, 117, 122–23, 125, 134, 136, 139–41, 143; whistleblowing and a, 72, 123. *See also* surveillance scenario
teaching, 2, 7, 15
technical communication: definition, 16, 47–50, 62; difference between technical communicator and Technical Communicator, 50; as a field, 3, 7, 15–16; scholarship, 16–17, 19, 21–22, 31, 33, 36–37, 59, 62, 64, 84–85
technical information, 15
technical writer, 15
technology: algorithms, code, and, 103, 135; and analytics, 93; and communication, 66, 68, 107; assess, 152; assisted surveillance, 41; creates categories, 102; data, information, and, 3, 16, 21–22, 66; demand for, 21; design, 74, 158; dystopian, 32; enabling surveillance, 18, 22; distributed, 159; feeling, affect, and, 36; and information communication, 22, 58, 67, 158; intersection of surveillance, communication, and, 16, 137; magazines, 137; media, 168n2; nonhuman actors and labor, 45, 135, 157; online, 77, 120; philosophy of, 158; platform, 15; police, technical communicators, and, 155; policy, laws, and, 18, 74, 156; privacy and, 143; program details, 66–68; provider, 78; to resist surveillance, 126; responsible, ethical, and empowering use of, 7; sites of surveillance, 18, 134, 137–38, 140, 151–52, 156n3; social impact and, 157; statistical discrimination and, 93; stop using to avoid being surveilled, 15; surveillance at a distance through,

27; surveillance and information, 21, 66–68; surveillance markets and, 22; surveillance embedded in, 10; and tactical communication, 120; technical communicator and, 15, 21, 47, 49, 52–54, 57–58, 60, 64–66, 120, 158, 163; technical communicator, expertise, and, 21; Tor, 67; types of, 65, 126; users, 16, 56, 58, 152; work-from-home, 92; worker, 42, 69

telecommuting, 13

television, 25, 31, 46, 162

Tempora, 100–1. *See also* projects of surveillance

terministic screen, 6

terms, 6

terrorism, 17, 93, 105–7, 112, 120

terrorists, 5, 7

threat: the aesthetic of a hacker to relay the, 26; of attack, 92; and de Certeau, 118; existential, 109; immigration as a, 95, 156; imminent 23; to research from surveillance, 17; insider, 57; ISIS, 107; as a potential to harm, 93, 105, 161; to risk taking behavior from surveillance, 20; security, 42 surveillance causes a, 2; of terrorists, 110–11; of violence from surveillance, 2, 29, 24

TikTok, 2, 145

Timan, Tjerk, 10, 13–14, 29

Tokyo, 44, 52

top-down, 14

tor, 67

track: behavior, 34; browsers, 101; cover one's, 43; customers, 135; foreign targets, 101; government will, 17; information, 135; laterally one another, 71; a period, 17; populations, 138; productivity, 135;

students, 19, 156; technology will, 10; terrorists, 111

traitor, 5

transparency: assurances of government, 109; as a basic liberty, 109; calls for more, 2, 103, 130; lack allows too much government privacy, 109; surveillance can bring, 33; is truth, 114; as a form of visibility, 109, 114, 162

transmit: data and information, 62, 75, 84, 101

Trottier, Daniel, 101

trust: diminished because of surveillance, 83; NSA, 110; newspapers, 106; positions of, 87

truth: co-construct, 25; conflicting, 88, 114; data analysis, 26–27, 43; and epistemology, 24–25, 93, 130; and rhetoric, 24–25, 34, 114, 130; and social media, 166n8; and surveillance, 25–27, 34, 93, 114; and technology, 27; and visibility, 25

ubiquitous: computing, 165n3; surveillance, 1, 10, 13, 137, 147, 165n3; technical communication, 60; technology, 16

ultrasounds, 17

unequal: distribution of goods and services because of surveillance, 28; distribution of power without transparency, 109; pardon, 112; pay because of surveillance, 28; treatment because of surveillance, 157

unfair: ethics, 99; power distribution, 109; surveillance, 37, 84

United States, 1, 5, 42, 44, 91, 106–7, 143, 154–56

United States Department of Justice, 110
United States Department of Labor, 42, 48, 52–55, 168n2
universal: norms, 97; statements about surveillance impossible, 23, 160; surveillance, 135
unwanted: gaze, 33, 98; surveillance, 117; visibility, 78
USA FREEDOM Act, 107–8, 110
usability, 36, 48, 75, 79, 139, 149
user, 11–12, 16, 31, 35–36, 43, 48, 54, 58, 64, 68, 74–75, 78–80, 92, 94, 101, 126, 128, 131, 139, 143–45, 152, 166n3
user experience (UX), 36, 51, 56, 58, 67, 70, 75, 79–81, 139, 143–44
utilitarian ethics, 95–97, 99, 103–5

value: as a benefit, 59, 129, 131, 143; commercial, 30; economic, 11, 96; people and, 30, 48, 89, 91–92, 96, 99; privacy, 97; of technical communication, 48; of work, 48
values: as in desirable for persons or society, 9, 28, 63, 86, 128; organizational, 128
Velasquez, Manuel G., 86–87, 95–100, 105, 108, 111, 113
violence, 2, 89, 91–92, 94, 121, 162. *See also* five faces of oppression
visibility: constant, 92; decrease, 117; disclosure and, 2, 129; Foucault, 9; increased, 77–78; information, 77; and otherness 156; power and, 39, 137; reality shows and, 25; social media and, 36; state, 94; surveillance and, 2, 6, 25–26, 30, 114, 117, 156, 161; systems of, 9. 37, 163; unwanted, 78
visible: documents make information more, 75, 78; harm, 29; make

surveillance 2, 8; make viewpoints more, 95, 114; resistance can be, 131
visual: appeal to audience, 63; design, 19, 48, 148; symbols of law enforcement, 26; symbols of surveillance, 26
voluntary: surveillance, 15, 71
voyeur: definition, 38
vulnerable: bodies, 129; groups, 37; populations, 129

Walwema, Josephine, 86, 119–21, 127
Walton, Rebecca, 29, 31, 37, 84, 86, 88–89, 91, 129, 148, 150, 155
Washington Post, 5, 101
watched, the, 14, 38, 46, 70, 93, 126, 128, 143
watcher, 8, 15, 38, 70, 74, 89, 93, 126, 143
watching: always has consequences and matters, 83; authority for, 134; behaviors, 9, 84, 143, 163–64; coworkers, 70; cultures, mistreatment, and 94; de Certeau and, 19, 128; has a disproportionate impact, 164; education and, 20, 156; employers and, 104, 163; everyday activities, 26; everyone is, 13; and ex-partners, 29; and fear, 35; for "truth" 25–27; Foucault and, 9–10, 13, 93; friends, 23; Google is, 35; indiscriminate, 23; is leveraged and used, 89; justice and, 137; legitimizes, 26; and lurkers, 23; and marginalized communities, 27; militarization and, 94; networked and distributed, 13; norms for, 23; online, 94; organizations, institutions, or corporations, 12, 23, 78–79; Orwell, Big Brother, and 32, 46; of others, 27, 79, 94, 135,

163; panoptic, 9; peers and, 35; physically, 10, 13, 158; platforms and, 35; pleasure, 84; power and, 8–9, 71, 89, 93, 161; puts order on the world, 92; regimes of, 31, 157; right to be, 6; self-censor from, 84; social media and, 45, 94, 166n8; someone is, 162; spies and, 23; state or government, 9, 13, 29, 32, 35, 46, 135, 156, 160; subtly, 163–64; suspects, 8, 27, 45; as a symbol of, 26; systems of, 19, 36, 92; technical communication and, 2; technical communicators and, 3, 20; technologies, 3, 126; threats, 23; TV, 25; under, 31; unequal, 157; visibility, 25; voyeur, 38; workers, 27, 149; you are, 13
wearables, 18
web browser, 13
web design, 15
website, 77–78, 109–10, 134, 139, 141–42, 169n2
Webster, William, 67
whistleblowing, 5, 71–72, 90, 104, 106, 118, 120–25, 127, 138, 140, 167n5, 168n1; definition, 122–24
wicked problem, 37
willing: agents and targets of surveillance, 134; to obey authority, 88; to participate in surveillance, 15, 71, 77; to provide information, 77; to sacrifice for exposure of surveillance, 109
Windows, 43, 53
Wise, J. Macgregor, 13, 32, 84, 165n1, 167n4
Wolk, Michaela, 161
women: surveillance's targeting of, 17
worker: benefit of the, 92; bodies of the, 92; case, 93; and the driving question of the book, 3, 159; everyday, 30, 135, 147; as a more general employee, 31, 42, 45, 49, 53, 58–59, 68, 91–92, 126–27, 138, 142, 158; government, 46; high profile, 25; information, 11, 21; law enforcement, 26; monitored, 92; non-information technology, 52; remote, 27; rights, 108; security, 43; surveillance and the, 2, 3, 6, 11, 37–38, 41–42, 44–47, 49–52, 55–56, 58–59, 61, 69, 81, 104, 142, 157, 159, 163, 167n1; technical, 6, 21, 38, 47; technical communicator and the, 69; technology, 42; UX, 80
Worker, Surveillance, 13, 44–47, 50–51, 55–59, 64, 69, 71, 121, 135, 149
workplace: espionage, 70, 79; ethics, 85; Foucault, 9, 19; general employment term, 49, 85, 99, 139; harassment, 19; *la perruque*, 126–27; as an omniscient organization that measures, 70; resistance, 126; rhetoric, 15, 33; surveillance definition, 70; as a surveillance site, 9, 16–17, 18, 33, 69, 71, 74, 76–79, 81, 92, 139; tactics, 126–27; technology, 18, 70, 76–78, 92; visibility, 9, 92
world leaders, 5
Wright, David, 120
writing center, 20

XKeyscore, 5, 101–2. *See also* projects of surveillance

Young, Iris Marion, 91
YouTube, 16, 31, 126, 150

zip code: as consumer category, 11, 34; high risk, 93; as surveillance "truth," 34, 93
Zuboff, Shoshana, 12, 14, 164n5

www.ingramcontent.com/pod-product-compliance
Lightning Source LLC
Chambersburg PA
CBHW030648230426
43665CB00011B/1008